D1451204

Plant Diversity

The Green World

Plant Diversity

J. Phil Gibson and Terri R. Gibson

Series Editor
William G. Hopkins
Professor Emeritus of Biology
University of Western Ontario

CHELSEA HOUSE
P U B L I S H E R S
An imprint of Infobase Publishing

Plant Diversity

Chelsea House
An imprint of Infobase Publishing
132 West 31st Street
New York NY 10001

ISBN 10: 0-7910-8960-6
ISBN 13: 978-0-7910-8960-6

Library of Congress Cataloging-in-Publication Data

Gibson, J. Phil.
  Plant diversity / J. Phil Gibson and Terri R. Gibson.
     p. cm. — (The green world)
Includes bibliographical references and index.
0-7910-8960-6 (hardcover)
1. Plant diversity—Juvenile literature. I. Gibson, Terri R. II. Title. III. Series.
QK46.5.D58G53 2006
580—dc22   2006023234

Text and cover design by Keith Trego and Ben Peterson

Printed in the United States of America

Bang IP 10 9 8 7 6 5 4 3 2 1

This book is printed on acid-free paper.

Table of Contents

Introduction

By William G. Hopkins

"Have you thanked a green plant today?" reads a popular bumper sticker. Indeed we should thank green plants for providing the food we eat, fiber for the clothing we wear, wood for building our houses, and the oxygen we breathe. Without plants, humans and other animals simply could not exist. Psychologists tell us that plants also provide a sense of well-being and peace of mind, which is why we preserve forested parks in our cities, surround our homes with gardens, and install plants and flowers in our homes and workplaces. Gifts of flowers are the most popular way to acknowledge weddings, funerals, and other events of passage. Gardening is one of the fastest growing hobbies in North America and the production of ornamental plants contributes billions of dollars annually to the economy.

Human history has been strongly influenced by plants. The rise of agriculture in the Fertile Crescent of Mesopotamia brought previously scattered hunter-gatherers together into villages. Ever since, the availability of land and water for cultivating plants has been a major factor in determining the location of human settlements. World exploration and discovery was driven by the search for herbs and spices. The cultivation of New World crops—sugar, cotton, and tobacco—was responsible for the introduction of slavery to America, the human and social consequences of which are still with us. The push westward by English colonists into the rich lands of the Ohio River valley in the mid-1700s was driven by the need to increase corn production and was a factor in precipitating the French and Indian War. The Irish potato famine in 1847 set in motion a wave of migration, mostly to North America, that would reduce the population of Ireland by half over the next 50 years.

As a young university instructor directing biology tutorials in a classroom that looked out over a wooded area, I would ask each group of students to look out the window and tell me what they saw. More often than not the question would be met with a blank, questioning look. Plants are so much a part of our environment and the fabric of our everyday lives that they rarely register in our conscious thought. Yet today, faced with disappearing rain forests, exploding population growth, urban sprawl, and concerns about climate change, the productive capacity of global agricultural and forestry ecosystems is put under increasing pressure. Understanding plants is even more essential as we attempt to build a sustainable environment for the future.

The Green World series opens doors to the world of plants. The series describes what plants are, what plants do, and where plants fit into the overall scheme of things. *Plant Diversity* introduces us to the seemingly endless variety in plant life, from the smallest alga to the tallest trees. We also learn the significance of plant diversity in planetary ecosystems and why understanding and protecting this diversity is critical to our own health and survival.

1 The Diversity of Plant Life

It is my pleasure here to take Botany as my special study,
which previously was the knowledge of a few plants;
today however the abundance of material for choice
has made it the most extensive of all the sciences.

—Carolus Linnaeus (1707–1778)
from the preface of *Species Plantarum*, 1753

The Diversity of Plant Life

Plants can be found nearly everywhere on Earth. They live in the cracks of New York City sidewalks. They survive in the sands of the Sahara. They thrive in the jungles of the Amazon basin. The plants themselves are no less diverse than their surroundings, coming in all different colors, sizes, shapes, and scents (Figure 1.1). Some plants, such as annual bluegrass (*Poa annua*) and pool sprite (*Amphianthus pusillus*), complete their life cycle in a matter of weeks, whereas individuals of other **species,** such as bristlecone pine (*Pinus longaeva*), can live hundreds or even thousands of years. These diverse features not only indicate the ability of plants to adapt to their environment, they also give clues to the evolutionary history of plants and insights into the history of life on Earth.

THE VALUE OF PLANT DIVERSITY

Knowledge and appreciation of plant **diversity** has always played a vital role in human survival. For the earliest hunter-gatherers, it was essential to know which plants were edible and which were not. Carelessly eating the wrong kind of leaf, fruit, or seed could cause sickness or even death. As cultures developed, early humans learned not only which plants to eat, but also which plants provided materials for housing, clothing, tools, and dyes. Plants such as sacred lotus (*Nelumbo nucifera*), peyote (*Lophophora williamsii*), and even cacao (*Theobroma cacao*), the plant from which chocolate is made, were used in religious ceremonies. Humans also learned which plants had medicinal properties. Individuals who knew which plants to use, where to find them, and how to prepare them were held in high regard as shamans, medicine men, or healers whose skill could mean the difference between life and death for their people.

Today, people still depend on plants for their survival. Plants make up a significant portion of the diet of the more than 6 billion people on this planet. Plants provide lumber and other construction materials for housing. They provide fibers for

Figure 1.1 Plants display a wide variety of forms. Photographed above are pine cones in a cluster *(a)*, ferns found in the Bialowieza Forest in Poland *(b)*, a cholla cactus *(c)*, and a bouquet of pink tulips *(d)*.

clothing and other items. Billions of dollars are spent annually on ornamental plants and **flowers**. Even the modern pharmaceutical industry continues to depend on plants, with more than 25% of prescription drugs containing compounds extracted from plants.

In addition to being useful to humans, plant diversity is also important to the functioning of planetary ecological systems. Plants direct the cycling of nutrients between the soil and other living organisms. Through **photosynthesis**, plants convert energy from the sun into other forms of energy that nonphotosynthetic organisms, such as humans and other animals, can use. Plants regulate temperature, influence precipitation patterns, provide habitat for other organisms, and perform many other essential ecological processes.

Unfortunately, many human activities, particularly habitat destruction, continue to have a negative impact on plant diversity worldwide. Development of effective conservation strategies to protect botanical resources will depend greatly upon knowledge of plant diversity and its ecological importance. Thus, appreciating and understanding plant diversity is critical for the survival and quality of all life on Earth.

Ethnobotany

Researchers estimate that more than 80% of the world's population still depends directly on plants for herbal medicines. **Ethnobotanists** are scientists who study the ways in which indigenous peoples use plants, particularly for medicinal purposes. Many botanists, ethnobotanists, anthropologists, and biochemists are currently working with native peoples around the world to learn about medicinal uses of different plants to preserve cultural traditions and knowledge. In the past, many medicines have been developed from plants used for healing purposes by tribal peoples.

THE STUDY OF BIOLOGICAL AND BOTANICAL DIVERSITY

The branch of biology that studies **biodiversity** is known as **systematics**. It is made up of three subdisciplines: **taxonomy, classification,** and **phylogeny.**

Taxonomy is the process of naming individual plants or groups of plants. A plant often has two names: a **common name** and a **scientific name.** Common names are given to plants by the people who live near them. Sunflower, loblolly pine, cottonwood, and crabapple are examples of common names frequently used by scientists and nonscientists alike. Common names are often easy to remember and are familiar to the general population.

Common names often describe some aspect of a plant's physical appearance. Names like pitcher plant, cat's claw, fish on a line, or leafy elephant foot conjure images familiar to everyone. Common names sometimes suggest a plant's uses (broom straw, match weed, scouring rush) or warn of its dangers (stinging nettle, poison ivy, death camas). However, this is not always the case. The names henbit, dogwood, and rattlesnake master are colorful, but give little information about the plant itself.

A plant may have more than one common name. For example, wax goldenweed and Spanish gold both refer to the same plant species, *Grindelia ciliata*. A common name may refer to several plants with similar characteristics. Aquacatillo ("little avocado"), for example, is a common name that identifies several different tropical tree species that produce small avocados. Consequently, common names can be a problem for scientists because they are not necessarily specific to a single plant, there are no rules to govern their application, and they provide limited information about a plant or its characteristics.

In the 1700s, a Swedish botanist named Carolus Linnaeus (1701–1778) reduced the **polynomial** (many name) system to a binomial system. Scientists currently use this **binomial nomenclature** (two name) system of scientific or Latin names to identify species. For example, the scientific name for the European grape is *Vitis vinifera*. The first part of the name (*Vitis*) is the **genus** and the second part (*vinifera*) is the species. A genus name can be used alone to refer to all members of a genus, but species names are never used alone.

The use of scientific names for plants is based upon a set of rules called the **International Code of Botanical Nomenclature** (**ICBN**). All names are in Latin or have been Latinized regardless of a plant's location, its possible uses, or any associations with a particular culture. Latin was chosen because this was the language of the classical works of botany and other sciences. Scientists around the world can communicate with certainty about a specific plant, regardless of the language they themselves speak. The following are some of the fundamental ICBN rules:

1. A plant or group of plants can have only one valid name.

2. The valid name for a species is the one first published closest to the date May 1, 1753 (the publication date for Carolus Linnaeus' book *Species Plantarum*, which is considered the starting point for modern taxonomy).

3. A name is valid if it has been published in scientific literature and contains a complete description of the plant written in Latin.

4. A valid name for a species consists of two parts, a genus name and a species name. These two parts may not be the same.

5. A genus name can be used only once, but a particular species name can be used in combination with different genera. For example, *Carya glabra* (pignut hickory) and *Rhus glabra* (smooth sumac) have the same species name *glabra* (meaning "without hairs"), but they are two different species.

6. The genus and species names are italicized or underlined.

7. The botanical rules of nomenclature are independent of the rules that govern naming of other organisms.

There are many other rules in the ICBN that give order to the process of naming. Botanists from around the world meet every

Families With Two Names

Although the ICBN states that there can be only one correct name for any group, there is an exception to this rule. Eight plant families have two correct, accepted names. The modern names have endings consistent with the ICBN rules. The older names deviate from ICBN rules, but have been in use for so many years (centuries, in some instances) that many botanists still use them today.

Table 1.1   Plant Families With Two Valid Scientific Names

ICBN name	Older name	Common name
Apiaceae	Umbelliferae	Parsley family
Arecaceae	Palmae	Palm family
Asteraceae	Compositae	Sunflower family
Brassicaceae	Cruciferae	Mustard family
Clusiaceae	Guttiferae	Garcinia family
Fabaceae	Leguminosae	Bean family
Lamiaceae	Labiatae	Mint family
Poaceae	Gramineae	Grass family

five years at the **International Botanical Congress** to discuss rules, propose changes, and make sure that the ICBN is serving the needs of the botanical community.

The second part of systematics, classification, is the process of grouping named organisms into an ordered system. Smaller groups are placed together into progressively larger groups, like a series of nesting boxes. In modern classification systems, the largest, most inclusive grouping is the **kingdom** (all plants are in the Kingdom Plantae). Plants are then divided into successively smaller groups, the smallest, and least inclusive of which is the species. A species is a particular kind of organism that typically can reproduce with other members of the same

Table 1.2   International Code of Botanical Nomenclature Endings

Taxonomic Level	ICBN ending	Scientific name
Kingdom	-ae	Plantae
Division	-phyta	Embryophyta
Class	-opsida	Angiospermopsida
Subclass	-idae	Asteridae
Order	-ales	Asterales
Family	-aceae	Asteraceae
Genus		*Grindelia*
Species		*Grindelia lanceolata*
Common Name		narrowleaf gumweed

Taxonomic Abbreviations

Systematists have developed their own terms and shorthand abbreviations to refer to different taxonomic groupings. One of the most commonly used words is **taxon** (pl. **taxa**), which refers to a group at any classification level. Genus names can be used alone or followed by the abbreviations sp. or spp., which are abbreviations for the word *species* (*sp.* refers to a single species and *spp.* refers to multiple species).

species and can be distinguished from other species based on a number of different characteristics. Members of the same species all share an evolutionary history distinct from that of other species.

The ICBN dictates specific endings for names at the different levels of plant classification (Table 1.2). For example, all plant family names end with the suffix *-aceae*. Orders, the classification level above family, end in *-ales*. Through this system, botanists can quickly recognize the taxonomic level of any name.

The final component of systematics is the study of phylogeny, the evolutionary relationship among organisms. Through this type of research, systematists can show how organisms evolved, how different organisms are evolutionarily related to one another, and how different groups have diverged and differentiated from one another throughout time. Studies of phylogeny often result in the construction of **dendrograms,** which are branching diagrams that show the evolutionary family tree for a group.

SIX KINGDOMS OF LIFE

Although there are differences of opinion, most biologists recognize six kingdoms: **Bacteria, Archaea, Protista, Animalia, Fungi,** and

Plantae. Organisms are placed into one of these kingdoms based upon their genetic and cellular features, as well as their modes of obtaining nutrients and energy.

Kingdoms Eubacteria and Archaea contain descendants of the oldest organisms on Earth. Members of these two kingdoms are **prokaryotic,** meaning that their cells lack a **nucleus** and membrane-bound **organelles** within them. Kingdom Bacteria is very diverse and contains a majority of the prokaryotic organisms on Earth. This group includes the true bacteria and cyanobacteria (blue-green algae). Kingdom Archaea is another diverse group. Many members of this kingdom often live in harsh environments, such as hot springs and deep sea thermal vents, in conditions similar to those found in Earth's early history.

The remaining four kingdoms are **eukaryotic,** meaning that their cells contain a nucleus and membrane-bound organelles. Kingdom Protista contains many unicellular organisms and simple multicellular organisms. Kingdom Protista contains organisms that share similarities with animals, fungi, and plants. Of particular importance are the green algae, which played an important role in the **evolution** of plants.

Kingdoms Animalia and Fungi contain multicellular, eukaryotic organisms that are **heterotrophs;** they must consume other organisms to obtain the nutrition and energy they need to live. Animals ingest the organisms they feed upon and secrete enzymes that break down their food into simple molecules, which are absorbed by tissues of the digestive system. In contrast, fungi secrete enzymes onto their food source, and then absorb the digested food from the environment. Fungi were long considered plants. Cellular and molecular data, however, have shown conclusively that fungi are not plants and should be placed in a separate kingdom. Interestingly, genetic data have shown that fungi are actually more closely related to animals than to plants.

Kingdom Plantae contains organisms that are **autotrophs,** which means that they are capable of producing their own energy. Botanists have identified more than 300,000 different species of plants; there may be as many as 500,000 different plant species on Earth. Plants are divided into four major groups: **bryophytes, seedless vascular plants, gymnosperms,** and **angiosperms.**

WHAT IS A PLANT?

Like all living things, plants are composed of cells, use energy and nutrients in their metabolism, and have evolved a variety of adaptations that help them survive in their environment. Plants also reproduce and interact with other living things. Several unique features, however, make it possible to group these organisms into their own kingdom.

A vast majority of plants are **photoautotrophs.** They conduct the biochemical process of photosynthesis in structures called **chloroplasts** (Figure 1.2). Through photosynthesis, plants convert energy in sunlight into chemical forms of energy, such as sugars and starch, that they can then use to meet their own needs and provide energy to other living things. Plants conduct photosynthesis using unique combinations of different pigments, such as **chlorophyll *a*, chlorophyll *b*,** and **carotenoids.** Some photosynthetic bacteria contain chlorophyll *b*, but only plants and green algae contain the other photosynthetic pigments. It is worth noting that there are parasitic plants, such as dwarf mistletoes (*Arceuthobium*) and Indian pipe (*Monotropa uniflora*), that do not produce photosynthetic pigments and are not photosynthetic; despite this, their biology and other traits clearly indicate that they are plants.

The plant cell wall is another unique feature of these organisms. Plant cells are surrounded by a wall made of complex carbohydrates, such as **cellulose** or **lignin,** which provides rigidity and structural support for the plant body.

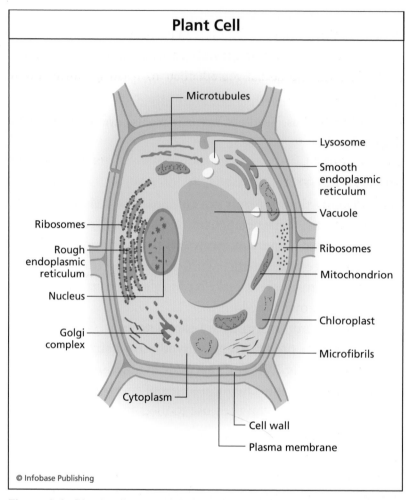

Figure 1.2 Plant cells are made out of many components, including a nucleus, cell wall, and chloroplasts.

Another defining trait of plants is that the structures in which **gametes** (eggs and sperm) are produced are surrounded by a layer of cells called the **sterile jacket,** which protects the developing gametes. Furthermore, the plant egg is not capable of movement and thus remains in female gametophyte tissues.

THE ALTERNATION OF GENERATIONS

All living things have a life cycle. The most familiar life cycle to most people is that of animals, in which single-celled gametes unite during **sexual reproduction** to form a multicellular **zygote** that grows and develops into a mature adult. Gametes are **haploid** (contain one complete set of **chromosomes**). Zygotes and adults are **diploid** (contain two complete sets of chromosomes). Haploid cells are produced when diploid cells undergo the process of **meiosis**, dividing twice to produce four haploid cells. **Mitosis** is a similar process, but instead of halving the genetic material in the cell, it produces two identical cells with the same amount of genetic material as the original parent cell.

Plants go through haploid and diploid phases in the life cycle. This unique type of life cycle is called the **alternation of generations** because during sexual reproduction there is an alternation between a multicellular haploid generation called the **gametophyte** and a multicellular diploid generation called the **sporophyte** (Figure 1.3). The gametophyte is a plant that produces gametes. Through **fertilization**, the egg and sperm unite to form a zygote that becomes the diploid sporophyte. The sporophyte is a plant that produces **spores**. The sporophyte grows through mitosis. Eventually, special cells in its reproductive tissues (**spore mother cells**) undergo meiosis to form haploid spores. The spores germinate to form the gametophyte, and the cycle repeats itself.

The alternation of generations is important in the study of plant diversity because the four major groups of plants differ with regard to which generation, gametophyte or sporophyte, dominates the life cycle. In bryophytes, for example, the gametophyte is the dominant generation, whereas the sporophyte is the dominant generation in angiosperms.

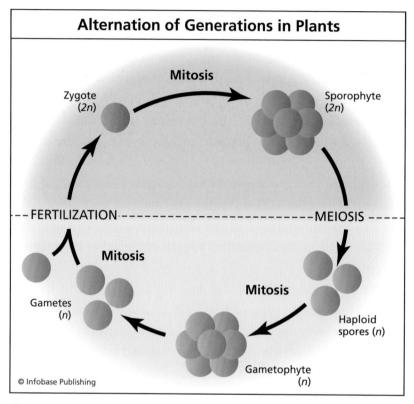

Alternation of Generations in Plants

Mitosis

Zygote (2n)

Sporophyte (2n)

FERTILIZATION ----------------------------------- MEIOSIS ----

Mitosis

Mitosis

Gametes (n)

Haploid spores (n)

Gametophyte (n)

© Infobase Publishing

Figure 1.3 Plants alternate between sporophyte and gametophyte phases. Sporophytes undergo meiosis to form spores, which initiates the gametophyte phase. Gametophytes have half the number of chromosomes as sporophytes. Male and female gametophytes, or gametes, combine during fertilization. The result of fertilization is the formation of a zygote.

SUMMARY

Plants are an important component of life on Earth. They provide a variety of ecological services ranging from the production of food through photosynthesis to providing habitats for other living things. Plants also provide a multitude of natural resources for humans and have performed important roles in the evolution of humans and the growth of human cultures and populations.

Through systematics, botanists identify and organize the diversity of plant life not only to catalogue botanical diversity, but also to understand its origins and evolutionary history.

2 The History of Plant Systematics

The botanist is he who can affix similar names to similar vegetables, and different names to different ones, so as to be intelligible to every one

—Carolus Linnaeus

The History of Plant Systematics

The earliest efforts to categorize plants were based on their usefulness to humans. Ancient texts and records describing medicinal, agricultural, and other uses of plants have survived from cultures around the world. One of the oldest such books was written more than 4,500 years ago by Emperor Chi'en Nung of China. One thousand years later, an unknown author wrote the Ebers papyrus, a scroll that described the pharmacological uses of plants found in ancient Egypt.

More than 2,300 years ago, the Greek naturalist Theophrastus wrote about the cultivation and uses of various plants, organizing them into groups based on characteristics such as their growth form (**herb**, **tree**, or **shrub**), **fruit**, and **leaves**. He gathered much of the information for his book *Historia Plantarum* (History of Plants) by studying the gardens of Athens. He eventually established the first known botanical garden.

Roman naturalist Pliny the Elder (A.D. 23–79) described the horticultural and medicinal uses of many plants in his book *Historia Naturalis* (Natural History). Dioscorides (circa A.D. 40–90), a Greek physician in Nero's army, wrote *De Materia Medica* (The Material of Medicine), a book that classified and described more than 600 plants. Not only did *De Materia Medica* contain written descriptions, it was also the first botany book to include illustrations. This book was one of the primary medical references in Europe until the 1600s.

One thousand years after Pliny and Dioscorides, a Chinese physician named Tang Shen-wei published *Cheng Lei Pen Ts'ao* (The Materials of Medicine Arranged According to Pattern). First published in 1108, this classic work of Eastern medicine went through 12 editions and was even translated and published in Japan in 1625.

THE REBIRTH OF BOTANY

In Europe, botany—like other sciences—lay mostly dormant until the Renaissance. The renewed interest at this time coincided with the invention of the printing press, which ushered in the so-called Age of Herbals. For the first time, leading German

Sedum minus flo: rubente.   Pœonia poly:anthos flore rubro.

Figure 2.1 Herbals are books that describe the many characteristics of plants used in herbal medicines. This illustration of a peony is from an herbal published in 1613.

botanists, such as Otto Brunfels, Hieronymus Bock, and Leonart Fuchs, were able to publish and widely distribute books that had descriptions and pictures of plants (Figure 2.1).

The first universities in Upper Italy and Europe established botanical gardens to study live plants. Luca Ghini (1490–1556), working at the University of Pisa, invented the process of

What Is an Herbarium?

An herbarium is a library of pressed and dried plants. In the herbarium, dried whole plants or parts of plants are mounted on pieces of paper called **herbarium sheets.** Each sheet has a label that identifies the plant and tells where and when the specimen was collected and by whom. Many professional systematists work at herbaria associated with botanical gardens, colleges, or universities. Specimens collected by Linnaeus are in herbarium collections around the world.

collecting, pressing, and drying plants to create the first **herbarium**, a library of preserved plant materials. The herbarium established in 1532 by Ghini's student Gherhards Cibo (1512–1597) is the oldest still in existence.

Although researchers continued to focus on the medicinal uses of plants, botanists of the seventeenth century began to investigate other areas, as well. Their discoveries led to the development of the first classification systems based on features of the plants rather than their uses. Naturalists, such as John Ray (1628–1705) and Pierre Magnol (1638–1715), described and named numerous European and Asian species, genera, and families. These botanists and others began to clarify concepts about plant families and genera that would become the foundation of modern systematics.

CAROLUS LINNAEUS

By the eighteenth century, botanists had collected, stored, studied, and classified a great deal of plant material from around the world. There was no standardized system or consensus, however, about how to name a plant—or any other kind of organism, for that matter. This led to a polynomial system that was cumbersome and impractical. For example, the full name for one plant species was *Serratula foliis ovato-oblingis accuminatus serratis, floribus corymbosis, calcybus subrotundis.* When the binomial system (see Chapter 1) was introduced by Carolus Linnaeus, this was shortened to *Serratula glauca.*

In *Species Plantarum,* published May 1, 1753, Linnaeus gave binomial names to more than 7,300 species (Figure 2.2). Although he was not the first person to consider a simplified binomial system, Linnaeus was the first to develop and consistently use a workable naming system—one that has been in use for more than 200 years.

NATURAL CLASSIFICATION SYSTEMS AND EVOLUTION

Although his binomial system was successful, Linnaeus's classification system was not widely accepted. Linnaeus categorized plants based on the number and arrangement of stamens, and then subdivided them using other floral traits. This arbitrary approach produced an **artificial classification system**, which grouped unrelated species together.

Contemporaries of Linnaeus, such as Michel Adanson (1727–1806) and Antoine Laurent de Jussieu (1748–1836), preferred to use a multitude of different features (for example, flowers, fruits, and leaves) to group plants based on overall similarity. Using this approach, De Jussieu described more than 100 plant families, many of which that are still recognized today.

Famous Botanists Remembered

Scientists often recognize individuals who have made major contributions to the field of botany by naming plants after them. The genus *Magnolia* is named after the botanist Pierre Magnol, whose work developed the concept of plant families. The genus *Dioscorea* (yams) is named in honor of the Greek botanist Dioscorides. Several paintings show Linnaeus wearing or holding a small cluster of *Linnaea* flowers, a genus that was named after him.

Figure 2.2 Carolus Linnaeus is known for his contributions to modern taxonomy. This famous picture of Linnaeus in Lapland dress shows him holding a cluster of flowers named after him, *Linnaea borealis*.

De Jussieu was one of the first botanists to group related species and families together. This approach produced a **natural classification system** that reflected evolutionary relationships among plants. The goal of producing natural classifications was further enhanced with the publication of *The Origin of Species* in 1859. In this groundbreaking book, Charles Darwin described how species evolve and how adaptations and other traits spread through the process of **natural selection**. Natural selection occurs when organisms have traits that help them survive better or produce more offspring than those that lack those traits. Through greater survival and offspring production, organisms with adaptive traits leave more descendants, and the adaptation becomes more prevalent in future generations. With his theory of evolution, Darwin showed that the diversity of life could be interpreted as a family tree, or genealogy, that reveals how taxa have changed throughout time.

EVOLUTION-BASED CLASSIFICATIONS

As the theory of evolution spread, botanists quickly developed evolution-based classifications. Prominent botanists, such as A. P. de Candolle (1778–1873) in Sweden; George Bentham (1800–1884) and Sir Joseph Hooker (1817–1911) in England; August Wilhelm Eichler (1839–1887), Adolf Engler (1844–1930), and Karl Prantl (1849–1893) in Germany; and Asa Gray (1810–1888) and Charles Bessey (1845–1915) in the United States, developed classification systems in which they considered the evolutionary history of specific traits. Some general rules they followed in developing evolution-based classifications included:

- Flowers with many parts (for example, many stamens or carpels) are less advanced than flowers with fewer parts.

- Plants with woody stems are more primitive than herbaceous plants.

- Species with flowers containing male and female structures are less advanced than species with flowers that contain only male or only female parts.

This new evolutionary perspective gave rise to modern classifications that show the phylogeny or evolutionary history and relationships among organisms in a lineage. Researchers continue to develop classifications that display the evolution of different plant groups.

MODERN CLASSIFICATIONS

Arthur Cronquist (1919–1992), Ledyard Stebbins (1906–2000), and Robert Thorne (b. 1920) of the United States; Armen Takhtajan (b. 1910) of the former Soviet Union; and Rolf Dahlgren (1932–1987) of Denmark produced major classification systems for angiosperms that influenced modern thinking on flowering plant evolution. Currently, systematists working in many different laboratories and herbaria take part in international collaborative efforts, such as the Angiosperm Phylogeny Group and the Tree of Life Project. These groups seek to clarify our understanding of the history of life on Earth by investigating large-scale patterns of plant evolution.

CLADISTICS AND CONSTRUCTING PHYLOGENIES

Systematists construct phylogenies using an approach called **cladistics**, a term that comes from the Greek word for branch, *klados.*

The Meaning of Theory

Evolution is sometimes discredited as a theory and not a fact. Scientists use the term *theory* to refer to a collection of unifying principles that explain proven facts and observations. Thus, the theory of evolution is not a guess, but a well-supported scientific explanation for the diversity of life.

Table 2.1   Comparison of Traits Between Algae and Major Green Plant Groups

Trait	Algae	Bryophytes	Seedless vascular plants	Gymno- sperms	Angio- sperms
Presence of chlorophylls *a* and *b*	yes	yes	yes	yes	yes
Sterile jacket layer around antheridia and archegonia	no	yes	yes	yes	yes
Vascular tissue	no	no	yes	yes	yes
Seeds	no	no	no	yes	yes
Flowers	no	no	no	no	yes

The objective of cladistics is to produce a natural classification in which every descendant of a single ancestor is placed on the same branch (**clade**) of a dendrogram. When all of the descendents of a common ancestor are placed in the same clade, it is called a **monophyletic group**. By identifying monophyletic groups, systematists can determine how specific traits in a lineage evolved throughout time.

For a cladistic study, systematists first determine which characteristics are primitive (ancestral) and then which are advanced. For example, Table 2.1 shows five different traits in the four major groups of plants and algae. The presence of a particular trait is a more evolutionarily advanced condition than the absence of a trait. For example, plants without vascular tissue

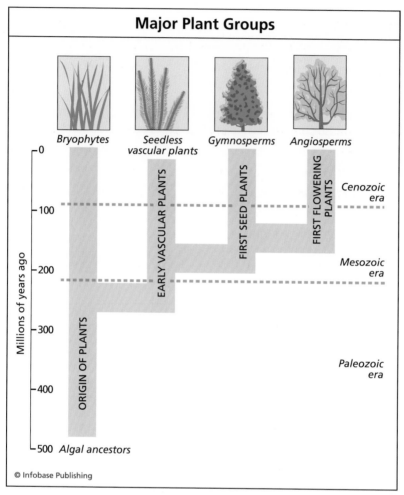

Figure 2.3 The above cladogram shows the relationship among major plant groups.

(**xylem** and **phloem**) appear in the fossil record before plants with vascular tissue. Therefore, the absence of vascular tissue in a group of plants is a more primitive condition than the presence of vascular tissue. Cladists use the patterns of how advanced traits arise to determine evolutionary relationships among members of a lineage.

The data in Table 2.1 can be used to produce a cladogram (Figure 2.3) that describes the phylogeny for the major groups

Botanical Resources

There are many resources available for individuals who want to explore the botanical diversity around them. One good resource is a **field guide.** These books often have pictures and descriptions that make it easy for budding botanists to learn the common plants of an area. More advanced students may require a **flora,** which typically contains an extensive listing of species that occur in an area. Field guides and floras typically contain **dichotomous keys,** which help users identify plants by leading them through a series of paired questions.

of plants. This cladogram shows that all plants shared a distant common ancestor with algae. Bryophytes are the oldest plant lineage, followed by the seedless vascular plants, gymnosperms, and ultimately the angiosperms. Systematists use the branching patterns of a cladogram to develop their classifications.

THE DATA SYSTEMATISTS USE

Systematics has been referred to as the field of science with no data of its own. In some ways, that is correct. Unlike the physiologist who collects data on the metabolic rates, cycles, and processes in cells, the geneticist who collects data on the sequence of genes on a chromosome, or the ecologist who collects data on the size and distribution of populations, systematists do not collect any form of data that one could specifically identify as systematics data. Instead, they use the techniques of other scientific fields to collect data that they can then use to develop a classification.

In the early stages of systematics, botanists depended heavily on **morphology** (the outward structural appearance of a plant). Traits such as leaf shape, fruit type, or number of parts in a flower are easy to observe and compare among living and fossil plants. With the invention of the microscope, botanists turned to **anatomy** (the internal cellular structure of a plant) to develop classifications.

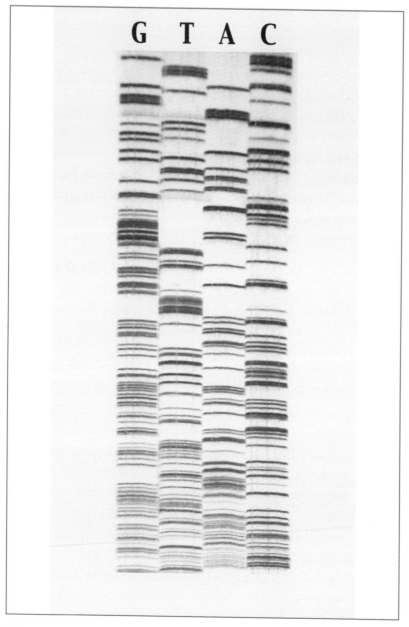

Figure 2.4 DNA sequencing gels are used by modern-day taxonomists to collect genetic data. Each band represents a single fragment of DNA known as a nucleotide. There are four types of nucleotides, abbreviated G (guanine), T (thymine), A (adenine), and C (cytosine).

The field of systematics is constantly changing as different technological advances arise. As instruments and techniques for analysis of plant chemical extracts improved in the 1940s, systematists began to collect chemical data for their studies. In the 1990s, there were tremendous advances in the ability to extract and sequence DNA from plants and improved computing ability to analyze data from DNA (Figure 2.4). Systematists rapidly incorporated these new techniques into their analyses. New techniques combine new data with earlier forms of data to further refine the understanding of plant evolution.

There is even continued discussion among plant systematists about how organisms should be named and classified. Recently, systematists proposed moving away from the Linnaean system toward a system called the **Phylocode,** which has no hierarchical rankings, such as family or order. Instead, it only recognizes clades. Although there is skepticism about the value of such a system and how it would affect the study of plant diversity overall, it does illustrate how botanists continue to think about naming, classifying, and interpreting the history of plant diversity.

SUMMARY

Throughout human history, individuals have worked to describe and organize the diversity of plant life on Earth. Early artificial classifications were based primarily upon a plant's usefulness to humans. Throughout time, systematists developed natural classifications that present the evolutionary history of plant groups. Linnaeus is one of the best-known names in the history of botany, but many different individuals have contributed and continue to play a significant part in this field of study.

3 Fungi and Algae

Did you hear about the fungus and the alga?
They took a lichen to each other.

—Unknown

Fungi and Algae

Although there are many beneficial species of fungi, people often fear them because of their toxins and other undesirable features. For example, fungi called molds cause allergies and other respiratory illnesses. Other species produce strong hallucinogenic compounds that have a range of effects on the nervous system. Other fungi, however, produce chemicals that benefit humans. The important antibiotic penicillin is derived from the fungus *Penicillium chrysogenum*. As a further example, the yeast *Saccharomyces cerevisiae* is an invaluable fungus used to make bread, beer, and wine.

Like so many of the fungi, algae too are often overlooked or unwanted. A thick growth of algae can pollute ponds and render water undrinkable (Figure 3.1). Fish tanks often need to be cleaned to remove algal growth. Other algae, however, are valued for their usefulness. Nori and sea lettuce are used in Asian cooking. Carrageenan, a product derived from algae, is used in ice cream, sauces, shampoos, cosmetics, air fresheners, and many other products. Algae are also of immeasurable ecological importance in oceans, lakes, and streams, where they are the foundation of the food chain.

At different times, fungi and algae have been classified as plants. Although these groups do share similarities with plants, fungi differ enough to warrant their own kingdom. Whether algae should be placed in a different kingdom, however, depends on which alga is being considered.

Fungi and algae have played vital roles in plant evolution. An ancestral algal species established itself in the terrestrial environment and gave rise to all plants. This invasion of land was made possible through interactions between plants and fungi. Because early plants had no roots, plants and fungi formed relationships in which fungi provided plants nutrients from the soil. A majority of plants today still form these important **symbiotic** relationships with fungi.

Figure 3.1 Excessive growth of algae can reduce the quality of freshwater lakes and ponds.

FUNGI

Mycology (from the Greek word for fungus, *mykes*) is the study of fungi. Like plants, fungi are multicellular eukaryotes whose cells produce a cell wall. Fungi cell walls, however, are composed predominantly of **chitin** (a substance also found in the external skeletons and shells of insects, crustaceans, and other related animals).

Fungi are composed of many small threadlike strands of cells called **hyphae.** The collection of hyphae that make the fungal

The Humongous Fungus

While several plants are candidates for the title of "largest or oldest organism on Earth," the fungi kingdom also has individuals of tremendous size and age. The honey mushroom (*Armillaria ostoyae*) is a fungus that attacks the roots of conifers. The mycelium of one individual in southwest Washington State was measured to cover 1,500 acres (approximately 2.5 square miles). Another, growing in eastern Oregon, covers 2,200 acres (3.4 square miles) and may be more than 2,400 years old.

body is called a **mycelium**. Fungi range in size from unicellular yeasts to the immensely large individuals that cover great areas in forests. Fungi are heterotrophs that feed on living and dead organisms. They obtain energy and nutrients by growing hyphae through the body of their food source and secreting enzymes that break it down into smaller, simpler, organic molecules, which are then absorbed. This method of feeding makes fungi, along with bacteria, extremely important as decomposers that clear organic waste from the environment and return nutrients to the soil.

BASIC FUNGAL LIFE CYCLE

Many fungi reproduce asexually as well as sexually. In **asexual reproduction**, spores are formed in a specialized structure called a **sporangium** (plural: sporangia) and then released to the environment. Fungi do not have male and female individuals. Instead, sexual reproduction occurs between genetically different **mating types.** For sexual reproduction, haploid mycelia of different mating types come into contact with one another, allowing cells from different hyphae to fuse. Although the cells combine, their nuclei can remain separate within them and grow as a mycelium composed of these **dikaryotic** cells. Whether the separate nuclei fuse immediately to form a zygote or continue to grow as a dikaryotic mycelium until forming a zygote at a later time is a distinguishing feature among different types of fungi.

MAJOR FUNGAL GROUPS

Mycologists have identified and named more than 50,000 species of fungi (Table 3.1 and Figure 3.2) and have estimated that the total number may be close to 1.5 million. Fungi are a monophyletic group, but their classification has been challenging. For example, recent advances in the collection and analysis of genetic and other molecular data have allowed mycologists to determine that the water molds (Oomycota) and slime molds (Dictyosteliomycota and Myxomycota), which had long been classified as fungi, are not fungi at all, but are in fact protists.

Presently, mycologists recognize five major phyla (equivalent to the botanical level of division) of fungi: chytrids (Chytridomycota), zygomycetes (Zygomycota), arbuscular mycorrhizal

Table 3.1   Major Groups of Fungi

Group	Common name	Estimated species
Chytridomycota	chytrids	1,000
Zygomycota	zygomycetes	1,100
Glomeromycota	arbuscular mycorrhizal fungi	157
Ascomycota	sac fungi	32,00
Basidiomycota	club fungi	26,000

Figure 3.2 Fungi display a wide range of growth forms. In the above photographs; two fly agaric mushrooms *(a)*, cultured *Beauveria bassiana (b)*, a bunch of mushrooms at the base of a withered plum tree *(c)*, and the fungal pathogen *Candida albicans*, as seen under a microscope *(d)*.

fungi (Glomeromycota), club fungi (Basidiomycota), and sac fungi (Ascomycota). Current perspectives on fungal evolution indicate that the chytrids and zygomycetes are the oldest lineages (Figure 3.3). Arbuscular mycorrhizal fungi, club fungi, and sac fungi are more advanced groups. Club fungi and sac fungi are closely related and encompass 95% of all known fungi.

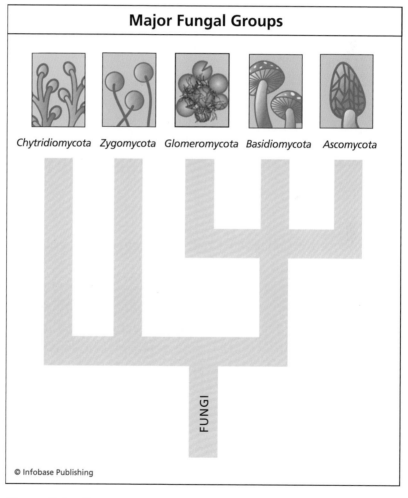

Figure 3.3   This cladogram shows the phylogeny of major fungal groups.

Chytrids

Chytridomycota are the oldest group of fungi. They live primarily in or near water and are the only group of fungi to produce spores and gametes with **flagella** (singular flagellum), a tail-like structure that gives cells the ability to move. The presence of flagella in this lineage provides evidence that, like all other organisms, fungi first evolved in the seas.

The chytrid life cycle is simple. Haploid spores germinate and form a haploid mycelium, which releases male and female gametes. The gametes fuse and immediately form a diploid mycelium that eventually forms a sporangium. Cells undergo meiosis in the sporangium and release the haploid spores.

Chytrids are important decomposers of plant material in many aquatic systems. A chytrid called *Batrachochytrium dendrobatidis* is currently attacking amphibian populations worldwide and may be responsible for the extinction of several amphibian species.

Zygomycetes

Zygomycota are a relatively small fungal group of about 1,100 species. A frequently encountered zygomycete is the common bread mold in the genus *Rhizopus*. In asexual reproduction, sporangia form on the tips of upright hyphae and release numerous spores. During sexual reproduction, fused hyphae from different mating types form a **zygospore** (the small black dots seen in moldy bread). A zygospore is highly resilient to environmental conditions and can remain dormant until environmental conditions cue favorable conditions for releasing spores.

Arbuscular Mycorrhizal Fungi

Mycologists recently moved a group out of Zygomycota and classified them as a new phylum, Glomeromycota. Their common name, arbuscular mycorrhizal fungi, highlights this group's important features. **Mycorrhizae** (which literally means "fungal root") are fungi that grow on a host plant's roots. The plant provides the fungus with a carbohydrate energy source. In return, the fungus provides the plant with nutrients from the soil. This type of plant-fungus **mutualism** occurs in the majority of plants on Earth. **Arbuscules** are specialized structures formed by the fungus inside the plant cell that enable the plant and fungus to exchange nutrients. The arbuscular mycorrhizal fungi are the only group of mycorrhizal fungi that form these specialized structures.

Arbuscular mycorrhizal fungi are obligate mutualists, meaning that they are unable to live without their plant host. Mycologists cannot even culture them in the laboratory. Sexual reproduction has never been observed in any member of this group.

Sac Fungi

Ascomycota is a large group that includes more than half of all known fungi. Their common name (sac fungi) comes from the sac-shaped structure called an **ascus** (plural: asci) that forms during sexual reproduction. Asci form on the edge of a large reproductive structure called an **ascocarp** that is characteristic of these fungi. During sexual reproduction, different mating strains

Chestnut Blight

American chestnut (*Castanea dentata*) once accounted for 25% of the trees in the eastern forests of North America. Then, in the early 1900s, a disease known as chestnut blight was accidentally introduced into North America from China or Japan. The disease, caused by the fungus *Cryphonectria parasitica,* attacked native chestnut trees and by 1940 had driven the species to the brink of extinction. In April 2006, however, a small group of chestnuts was discovered near Warm Springs, Georgia. Scientists believe that these trees survived because the site is too dry for the fungus.

fuse, forming a dikaryotic mycelium. The mycelium then forms an ascocarp with the asci on it. Inside the asci, the nuclei fuse, and then immediately undergo meiosis to form haploid spores, which germinate to establish a new mycelium.

Ascomycetes are important decomposers in many ecosystems. The sac fungi also include many species that cause disease in a wide variety of crops and wild species. There are also many beneficial ascomycetes. The genus *Penicillium* includes the source of penicillin (*Penicillium chrysogenum*), the blue cheeses (*Penicillium roqueforti* and *Penicillium camembertii*), and the yeast used in brewing and baking (*Saccharomyces cerevisiae*). Morels (*Morchella*) and truffles (*Tuber*) are considered delicacies by some individuals.

Club Fungi

The mushrooms, toadstools, and puffballs common to forests and front yards are members of the Basidiomycota. The name *club fungi* refers to the club-shaped sporangium, called a **basidium** (plural: basidia), formed during sexual reproduction. As in other groups of fungi, the sexual reproduction of basidiomycetes begins with the joining of dissimilar mating strains. The resulting dikaryotic mycelium forms a reproductive structure called a **basidiocarp** (the familiar mushroom). The **gills** on the undersurface of the basidiocarp form numerous basidia in which the dikaryotic nuclei fuse to form a zygote. The single-celled zygote then undergoes meiosis to form haploid spores that are released to germinate and produce new mycelia. The zygote is the only diploid cell in the entire basidiomycete life cycle.

Club fungi are a diverse group that has many different properties and uses. The white button or portabella mushroom (*Agaricus bisporus*), oyster mushroom (*Pleurotus ostreatus*), and shiitake mushroom (*Lentinula edodes*) are eaten on salads, pizzas, and many other dishes. *Amanita bisporigera*, *Amanita virosa*, and *Amanita verna* are known as death angel fungi because of

Figure 3.4 Smut, a fungus that causes plant disease, attacks many cereal crops. Above, mature smut galls are exposed on an ear of corn.

their lethal toxins and almost pure white color. Fly agaric (*Amanita muscaria*) and psilocybin mushrooms (*Psilocybe cubensis*) produce hallucinogenic compounds that have been used by shamans—a religious specialist or medicine man—in ancient cultures and religions worldwide. Shelf fungi are important decomposers of wood and other plant materials. Club fungi called rusts and smuts are serious plant pathogens that cause billions of dollars in crop damage annually (Figure 3.4).

ALGAE

Phycology, the study of algae, covers a wide range of photosynthetic organisms—from minute diatoms, whose shells display intricate patterns and structural variations, to the enormous kelps that grow in the cold oceans of northern latitudes. This diversity has allowed algae to succeed in a wide range of habitats. Algae live in soil, sand, and on the barks of trees. Some even grow in snow. Many live in the oceans, where they form the base of marine **food webs**. Others perform the same ecological role in freshwater ecosystems. Some green algae have even evolved unusual associations with other organisms.

Green algae are of particular importance because they are the group from which land plants evolved. Two other groups—the red algae (division Rhodophyta) and the brown algae (family Phaeophyceae)—are also included here, however, because of their importance and familiarity to humans.

Green Algae

Green algae are ancient; their origins can be traced back more than 900 million years. Throughout the course of time, the one original unicellular species diversified to form other unicellular species, as well as larger, more complex, multicellular growth forms.

Green Alga Life Cycle

Green algal life cycles include sexual and asexual reproduction. In unicellular algae, asexual reproduction via spore formation or cell division is common. Sexual reproduction in unicellular green alga involves flagellated, haploid individuals of different mating types simply coming together and fusing nuclei. The resulting diploid zygospore undergoes meiosis and releases haploid unicellular individuals. Cells are also typically haploid in multicellular green algae. During sexual reproduction in multicellular species, some of the haploid cells release flagellated,

It's Not Easy Being Green

Algae are well known for their mutualistic relationships with fungi. Less well known, however, are their mutualistic relationships with animals. One such relationship involves the sloth, a mammal that lives in the canopy of tropical forests. The sloth gets its greenish hue from the algae that live in its fur. This coloration helps camouflage the sloth, protecting it from predators. In return, the algae get a place to live and exposure to the light that is present high in the forest canopy, where sloths usually dwell.

unicellular gametes, which join in a process similar to that of unicellular algae. The resulting single-celled zygote undergoes meiosis and releases haploid spores that grow to become multicellular individuals.

Green Algae Lineages Are Related to Land Plants

Green algae are a large group with more than 17,000 species. There are three major groups within the green algae: prasinophytes, chlorophytes, and the charophytes. Prasinophytes are a diverse group of unicellular marine algae. This group includes several separate lineages and, thus, is not a monophyletic group. Prasinophytes are important because they make up some of the oldest lineages of green algae and are therefore representative of the first green algae to evolve.

Chlorophytes are a group of approximately 7,500 species of green algae that live in salt water, fresh water, and on land. They exhibit a number of different growth forms ranging from single-celled species (*Chlamydomonas*), filamentous forms (like the genus *Oedogonium*), spheroid colonies (such as *Volvox*), and the delicate, sheet-like sea lettuces (*Ulva*).

Charophytes are predominantly freshwater or terrestrial algae, although some live in brackish water (a combination of salt and fresh water). Charophytes, like chlorophytes, include a diversity of unicellular and multicellular growth forms. Some charophytes in the order Charales are called stoneworts because of calcium deposits in their cell walls, a trait that has helped preserve evidence of them in the fossil record. The genus *Spirogyra* contains many filamentous species commonly encountered in freshwater lakes and ponds.

Relationship Between Green Algae and Plants

Green algae share a number of traits with plants, but they also differ in important ways. Algae lack a sterile jacket layer around their gamete-producing structures. In addition, a vast majority of algae release their gametes into the water, whereas the female gamete is retained on the maternal individual in plants.

Charophytes are the only organisms other than plants that produce pigments called **flavonoids**. Species in the genera *Chara* and *Choleochaete* also have reproductive traits, such as flagellated sperm and nonmotile eggs, that are similar to terrestrial plants. Species in these groups also retain the zygote in the gametophyte tissues, but they lack the sterile jacket cells found in plants. Systematists believe that algae in the *Chara* lineage are the living descendents of a common ancestor shared by green algae and land plants. Although systematists agree that land plants evolved from a green alga ancestor, there is continuing discussion about whether plants and green algae should be grouped together or separately.

Red Algae (Rhodophyta) and Brown Algae (Phaeophyceae)

The debate about the exact relationship between green algae and modern plants does not extend to the red and brown algae. These groups differ from plants in several ways. Because they share structural features with the green algae, however, and because they are photosynthetic autotrophs, they are included in this discussion of groups that are related to land plants.

Brown algae are a small group of approximately 1,500 species that live mostly along the rocky shores of cold ocean waters. Kelp is an important brown alga that forms extensive underwater "forests." The kelp body has three specialized regions: a **holdfast** that anchors the kelp, a **stipe** (or stem), and **blades** (leaves). Blades often have a **float,** which is an air-filled structure that holds the kelp upright. Among other things, brown algae differ from plants in the photosynthetic pigments they contain.

Red algae are a large group of approximately 6,000 marine species, most of which are multicellular. They get their characteristic color from the high amount of reddish pigments called **phycobilins** in their cells. These pigments are particularly good at absorbing what little light there is in the deepwater environments where many red algae species live. The deepest known photosynthetic organism is a red alga that has been found living 268 meters below the water's surface. Often, the cells of red algae are covered by a jellylike substance (mucilage) or contain deposits of a certain type of mineral (calcium carbonate); these materials help the red algae survive in deep water. Although some red algae produce toxic substances, they do not cause the red tides that kill fish and other marine life. Red tides are actually caused by population explosions of protists known as dinoflagellates.

LICHENS

Lichens are interesting organisms that develop from a mutualism between fungi and algae or cyanobacteria. More than 98% of the fungi that form lichens are ascomycetes, but a few basidiomycetes also form lichens.

In lichens, many unicellular algae live embedded in a fungus (Figure 3.5). The algae provide carbohydrates and other nutrients to the fungus, which in turn provides a livable environment for the algae. Lichens exist in extremely stressful environments, such as tree branches or bare rock, in which neither the fungus

Figure 3.5 Lichens are symbiotic organisms with algae embedded in a fungal matrix. The above image is of *Lobaria pulmonaria,* the most widespread *Lobaria* lichen, photographed in Oregon.

nor algae could live alone. This ability to survive in harsh environments allows lichens to be among the first colonizers of bare rock surfaces.

Lichens display three different growth forms. **Crustose** lichens are flat; they often appear to be painted on the surfaces where they grow. **Foliose** lichens have a leafy appearance. **Fruticose** lichens range in appearance from the multibranched individuals of reindeer moss (*Cladonia subtenuis*) to the erect pillars of British soldier (*Cladonia cristatella*). Despite their ability to tolerate the hot, dry, bright conditions of bare rock, lichens are quite sensitive to air pollution. Environmental scientists therefore use lichens as indicators of air pollution.

SUMMARY

Although they are not plants, fungi and algae are important in any discussion of plant diversity and evolution. Fungi are multicellular, heterotrophic organisms. Their ecological function as decomposers makes nutrients in the soil available to plants. Symbiotic relationships with mycorrhizal fungi allowed early plants to colonize the land, and continue to help them still. Green algae are a diverse group of unicellular and multicellular photosynthetic organisms whose characteristics indicate that all terrestrial plants evolved from a single algal ancestor.

4 Seedless Nonvascular Plants
The Bryophytes

... the peat of the valley of the Somme is a formation which,
in all likelihood, took thousands of years for its growth.

—Charles Lyell (1797–1875)
British geologist

Seedless Nonvascular Plants
The Bryophytes

More than 500 million years ago, a multicellular species of alga became the first organism to make the transition from living in the water to living on land. Living in shallow waters, this ancestor of all plants experienced alternately wet and dry conditions with the rise and fall of the tides. Traits evolved in this species that allowed it not only to tolerate the dry periods at low tide, but also to survive farther from the edge of the water. From these simple beginnings, Kingdom Plantae was born, and the botanical invasion of land had begun.

Plants adapting to land had to overcome certain challenges; air does not provide the structural support or the nutrients that water does. In response to these new environmental conditions, early plants began producing a **cuticle**—a thickened layer of waxy material on the outer surface of cells that prevented them from drying out. The plants were probably thin, allowing nutrients to enter and spread throughout the plant body (**vascular tissue** for specialized transport would not evolve for another 50 to 100 million years).

Although often overlooked because of their diminutive size, bryophytes (Figure 4.1) are the living descendents of these earliest plants. Bryophytes are nonvascular plants. There are between 20,000 and 25,000 species of bryophytes (Table 4.1) classified into three divisions: liverworts (Hepatophyta), hornworts (Anthocerotophyta) and true mosses (Bryophyta). These modern bryophytes have retained many features of their ancestors; for example, they produce no vascular tissue or seeds. Because of this, the study of bryophytes can provide insight into some of the first strategies evolved by plants in their adaptation to life on land.

FEATURES OF BRYOPHYTES

Bryophytes are typically small plants, ranging from a few millimeters to several centimeters in height. Their size is limited because they lack vascular tissue, which provides the structural support and nutrient transport necessary for larger growth. Instead of

Figure 4.1 Bryophytes are small, nonvascular plants. Above, green moss (a type of bryophyte) covers three oak tree trunks.

Table 4.1   **Major Groups of Bryophytes**

Group	Common name	Estimated species
Anthocerotophyta	hornworts	100–150
Hepatophyta	liverworts	6,000–9,000
Bryophyta	true mosses	12,000–15,000

using roots to take up water and anchor themselves in place (as plants do), bryophytes use specialized cells called **rhizoids**. Unlike root cells, rhizoids do not have any greater water uptake ability than the rest of the plant. Most bryophytes use their entire bodies to gather water and nutrients from the environment.

Although some bryophytes can survive in very dry habitats, a vast majority of bryophytes live in moist environments (Figure 4.2). Bryophytes are restricted to moist environments partly because they lack vascular tissue, but also because water is required for sperm to swim to the eggs during sexual reproduction.

BRYOPHYTE LIFE CYCLE

The life cycle of bryophytes has distinct gametophyte and sporophyte generations. The free-living gametophyte is the dominant phase. The short-lived sporophyte is attached to and nutritionally dependent upon the gametophyte. The structure and appearance of gametophytes and sporophytes differ among bryophyte groups.

Haploid spores germinate into small, threadlike, multicellular gametophytes called **protonemas**. These protonemas mature and produce gametophytes with **antheridia** (sperm-producing structures), **archegonia** (egg-producing structures), or both. Most bryophytes produce antheridia and archegonia on the same

Figure 4.2 Mosses are common in moist, streamside habitats. In this photograph, moss is present on a boulder near Silverton, Colorado.

gametophyte, but mosses produce antheridia and archegonia on separate male and female gametophytes.

Because the sperm must swim to the egg, male and female gametophytes must grow near one another. Sperm cells use whiplike flagella to propel them down a narrow canal toward the egg. After fertilization, the combined egg and sperm form a diploid embryo that will become the sporophyte. The young sporophyte is attached to the gametophyte and will remain attached to the gametophyte for its entire life. A sporangium forms at the tip of the sporophyte. Spore mother cells inside the sporangium will undergo meiosis to form the haploid spores that will start the cycle over again.

MAJOR GROUPS OF BRYOPHYTES

Bryophytes are a relatively small group of plants containing approximately 24,000 species. All bryophytes are structurally simple plants, and the major groups differ primarily in their reproductive structures.

Figure 4.3 *Marchantia* is a typical liverwort. It is a primitive plant related to mosses and ferns.

Doctrine of Signatures

Liverworts and an assortment of other plants include the names of various organs in their common names. This comes from a belief known as the Doctrine of Signatures. Early physicians and herbalists thought a plant that resembled an organ could be used to treat ailments of that organ. The lobed liverwort thallus was used to treat liver complaints because it resembles a human liver. The suffix *-wort* comes from the Old English word *wyrt,* meaning "herb." The Doctrine of Signatures was a popular medical concept through the nineteenth century and is still used by some practitioners of homeopathic medicine.

Liverworts

Liverworts are the second largest bryophyte division, with 6,000–9,000 species. The basic body form of a liverwort consists of a simple, flattened, leaflike **thallus,** that has a very simple branching pattern (Figure 4.3). These plants are in a clade called the leafy or simple thallus liverworts. The remainder of the liverworts are in a clade called the complex thallus liverworts. These plants have a broader, flattened thallus with pronounced structural differences between their upper and lower layers.

The genus *Marchantia* contains species common to cool, shady, moist areas near streams and waterfalls. *Marchantia* produces separate male and female gametophytes. The males produce antheridia on raised flattened structures which, when hit with a drop of water, launch the sperm from the gametophye. Females produce archegonia on a raised structure that has a drooping appearance. The sporophyte is a small, rounded structure attached to the underside of the female gametophyte. **Gemmae** (singular: gemma) are small masses of tissue that may form in cuplike structures on the thallus. These masses of vegetative cells can splash out of the cups, allowing the parent plant to reproduce asexually through **vegetative reproduction.** *Marchantia* may also reproduce vegetatively by fragmentation of parts from a larger, older plant body.

Figure 4.4 Hornworts are most commonly found in damp, humid, undisturbed locations. Hornworts are seen here in Lincoln County, Oregon.

Hornworts

Hornworts are the smallest bryophyte division, with 12 genera and only 100–150 species. Like the liverworts, they produce antheridia and archegonia on the upper surface of a thallus. In hornworts, however, the sporophyte grows from the upper surface of the thallus and has an elongated horn or spindle shape (Figure 4.4). Hornworts resemble vascular plants in that their sporophyte is photosynthetic and produces **stomata** (openings that allow water and gasses to be exchanged between the plant and the atmosphere). The production of stomata may be an important evolutionary link between hornworts and plants that evolved later.

Mosses

The most familiar of the bryophytes, mosses, are a very successful group, with 12,000–15,000 species. They grow in a wide range of conditions ranging from moist areas along streams and in forests to arctic and alpine environments. Mosses are typically divided into three classes: peat mosses (Sphagnopsida), granite mosses (Andreaeopsida), and true mosses (Bryopsida). The peat mosses are a small group with two genera and more than 400 species. *Sphagnum* peat bogs cover approximately 1% of the Earth's surface. Granite mosses are an even smaller group of approximately 100 species. These blackish-green or reddish-brown mosses grow in clumps in mountains and granite outcrops. The true mosses are the most common. Although individual plants are small, moss gametophytes can dominate the landscape with extensive mats of vegetation.

The moss sporophyte grows from the top of the female gametophyte. The mature sporophyte is composed of a capsule on the tip of a long stalk called a **seta**. When mature, the capsule dries, releasing the numerous spores inside.

Similar to vascular plants, mosses produce specialized transport cells. Cells called **hydroids** are found in the central stem of some moss species, and function to transport water up the moss stem. Cells called **leptoids** transport sugars. Although they function like vascular tissue, hydroids and leptoids lack some of the defining features of true vascular tissue. Many mosses have a leafy appearance with areas of photosynthetic tissue that resemble leaves. Because these structures lack vascular tissue, however, they are not true leaves. Nevertheless, such features indicate a close relationship between mosses and vascular plants.

A Moss By Any Other Name

Confusion between common and scientific names is nowhere more evident than in the term *moss.* Technically, mosses are the plants in the group Bryophyta. The name *moss,* however, is applied to various plants and nonplants outside of this group. Irish moss and the moss in ponds and lakes are actually algae. Spanish moss and rose moss are angiosperms. Club moss and spike moss are seedless vascular plants related to ferns. Reindeer moss is not a plant at all, but rather a lichen.

Table 4.2   The Geological Timescale

Eras	Periods	Beginning (mya)*	Major botanical events
	Quarternary	1.8	
Cenozoic	Tertiary	65	Radiation of flowering plants
	Cretaceous	144	Evolution of flowering plants
Mesozoic	Jurassic	206	
	Triassic	248	Conifers dominant
	Permian	290	
	Carboniferous	354	Forests of large, primitive trees
Paleozoic	Devonian	417	Seedless vascular plants dominate, first seed plants
	Silurian	443	First vascular plants
	Ordovician	490	Plants invade land
	Cambrian	543	
Precambrian		4550	Formation of Earth

*mya: million years ago

ORIGINS, DIVERSIFICATION, AND PHYLOGENY OF THE BRYOPHYTES

Spores and fragments of ancient bryophytes have been found in fossils that date back almost 445 million years (Table 4.2). These fossils and other data suggest that bryophytes evolved from their algal ancestor between 490 and 500 million years ago. Bryophytes probably diverged from the lineage that gave rise to vascular plants about 430 million years ago. Fossils of ancient liverworts have been found in rocks dating back 360 to 408 million years, and moss fossils have been identified from as far back as 290 to 360 million years ago.

Until recently, all bryophytes were thought to have descended from a single, common ancestor. New evidence indicates, however, that bryophytes are actually three distinct monophyletic lineages. The arrangement of the three lineages, however, is not completely certain. Traditionally, systematists believed hornworts to be oldest, with liverworts the next oldest, and mosses the most recently evolved group of bryophytes. Some studies, however, have suggested that hornworts may actually be the most closely related to vascular plants.

ECOLOGICAL AND ECONOMIC IMPORTANCE

Ecologically, bryophytes are important plants. They grow in extensive mats and colonies that can help stabilize soils near streams. They may colonize disturbed sites early, beginning the process of ecological **succession** in which the plants growing on a site changes over time. Mosses are important components of food chains in many arctic and alpine environments.

Economically, bryophytes are of some importance. *Sphagnum* moss has a variety of different uses. In the past, it was used to dress and pack wounds. Peat, which is largely *Sphagnum*, is burned as fuel in some northern temperate areas, in homes and in larger-scale electricity production. Peat is also important in the horticulture industry where it is used to lighten and increase the water-holding capacity of soil.

SUMMARY

Liverworts, hornworts, and mosses are nonvascular, nonseed-producing plants. Known collectively as bryophytes, these were the first plants to successfully invade, establish, and persist on land. True to their aquatic origins, however, they still depend on water to transport sperm to egg during sexual reproduction. Bryophytes are the only plants with a dominant gametophyte stage and a smaller sporophyte stage.

5 Seedless Vascular Plants

Nature does not proceed by leaps and bounds.
—Carolus Linnaeus

Seedless Vascular Plants

The coal that powered the Industrial Revolution in the eighteenth and nineteenth centuries is still an important fuel source today. Coal formation began more than 300 million years ago, when shallow seas covered the Earth and a warm tropical climate promoted lush plant growth. Trees and plants that fell into the water as they died did not decompose because the low oxygen content of the water prevented it. Over time, layers of sediments accumulated on top of the plants, which forced the water out of their tissues. During the course of millions of years, the intense pressure and heat of geologic forces converted the plants into coal.

One genus of tree, in particular, contributed a great deal of plant matter to the coal beds. The trunks of *Lepidodendron* trees stood more than 130 feet (40 meters) tall and were more than 6 feet (2 meters) in diameter. They and their relatives formed extensive forests. *Lepidodendron* and many of the other plants that eventually became coal were members of a group called the seedless vascular plants. Unlike the bryophytes before them, these plants evolved vascular tissue, a network of xylem and phloem that transports water, nutrients, and other materials efficiently and effectively around the plant body. This and other important traits made the next great breakthrough in plant evolution possible.

Many different lineages of seedless vascular plants have become extinct since they first evolved. At present, seedless vascular plants consist of approximately 14,000 species in two monophyletic clades (Table 5.1). The oldest clade, the lycophytes (division Lycopodiophyta), contains three orders: the club mosses (Lycopodiales), spike mosses (Selaginellales), and quillworts (Isoetales). The second clade, previously known as the ferns and fern allies, is the monilophytes. This clade contains horsetails (Equisetales), whisk ferns (Psilotales), ophioglossoid ferns (Ophioglossales), marattioid ferns (Marattiales), water ferns (Salviniales and Marsileales), and the true ferns (Filicales). The monilophytes shared a common ancestor with the lineage that eventually gave rise to seed plants.

Table 5.1   Major Groups of Seedless Vascular Plants

Group	Common name	Estimated species
Lycopodiales	club mosses	380–400
Selaginellales	spike mosses	700–750
Isoetales	quillworts	125
Psilotales	whisk ferns	15–17
Equisitales	horsetails	15–25
Ohioglossales	ophioglossoid ferns	75–110
Marattiales	marattoid ferns	200–240
Marsiliales	water clover	75
Salviniales	water spangles, mosquito ferns	16
Filicales	true ferns	11,500–12,000

FEATURES OF SEEDLESS VASCULAR PLANTS

Seedless vascular plants were the first to evolve vascular tissue. These plants, along with the gymnosperms and angiosperms, form a group called the **tracheophytes** or vascular plants. The most primitive type of water-conducting cells, **tracheids,** are found in the xylem of seedless vascular plants. These cells are rigid due to the presence of lignin in their cell walls. Networks of tracheids not only aid the passage of water, but also support the plant body. This internal reinforcement of the stem allowed plants of the Carboniferous to reach greater heights than ever before. Eventually, plants evolved the mechanisms to produce and accumulate xylem in the stem, which led to the evolution of wood and even stronger stems and branches.

With vascular tissue, came the evolution of leaves. Leaves are thin, flattened areas of photosynthetic tissue surrounding a network of veins. Leaves evolved from highly branched stems of early vascular plants, when photosynthetic tissue filled in the spaces between the finest branches at the tips of the stems. Vascular plant leaves and stems have a thick cuticle layer on their outer surface that prevents water loss. Leaves also became much more elaborate and structurally diverse in seedless vascular plants. The ferns in particular evolved large leaves called **fronds** that produce sporangia on their underside in many species (Figure 5.1).

Plants were evolving new structures below ground as well. Instead of the simple rhizoids that attach bryophytes to the ground, the first vascular plants evolved roots. Larger roots enabled plants to extract water and nutrients from the soil more effectively. Larger roots also provided a stronger anchor and better structural support to the larger stems that were evolving above ground.

The life cycle of seedless vascular plants also diverged from that of their bryophyte ancestors. While the sporophyte in bryophytes is small and completely dependant upon the gametophyte, the gametophyte and the sporophyte in the seedless vascular plants can be a free-living plant. Furthermore, unlike the brypophytes, the sporophyte dominates the life cycle in seedless vascular plants.

These changes in the vascular tissue, leaves, and sporophyte promoted the success of the seedless vascular plants. Vascular tissue and leaves helped plants survive in a terrestrial environment and allowed them to colonize drier areas on land. The dominant sporophyte also allowed the plants to produce far more offspring than bryophytes.

Figure 5.1 Sporangia are found on the underside of fern fronds. Sporangia contain spores, and when the walls of the sporangium dry out, the spores catapult away from the plant.

SEEDLESS VASCULAR PLANT LIFE CYCLE

The life cycle of a typical fern serves as a good example of the alternation of generations in seedless vascular plants. Spores are released from sporangia on the undersurface of the leaves. The spores germinate to form an independent, heart-shaped, photosynthetic gametophyte. At maturity, the antheridia release the sperm, which swim to and fertilize the egg housed in the archegonium. After fertilization, the embryonic sporophyte grows from the gametophyte, which then withers and dies while the sporophyte becomes the plant we know as a fern.

Another significant change involving the spores occurred in the seedless vascular plants. In the bryophytes and earliest vascular plants, all spores produced by a plant are identical (**homospory**), producing gametophytes with antheridia and archegonia (mosses have separate male and female gametophytes, but they come from identical spores). Some vascular plants evolved the ability to produce different spores (**heterospory**). **Microspores,** produced in a **microsporangium**, give rise to male gametophytes. **Megaspores,** produced in a **megasporangium,** make the female gametophyte. Heterospory is a valuable trait because it allows some gametophytes to become specialized for sperm production and others for egg production. This was an important step toward the evolution of **pollen** and **seeds** that would occur later in the gymnosperms.

MAJOR GROUPS OF SEEDLESS VASCULAR PLANTS

Seedless vascular plants include a diverse array of more than 25,000 species. Members of the major groups display a wide range of growth forms, life cycles, and other adaptations to the environments where they live.

Lycophytes

Lycophytes are a relatively small group of plants with 3 families, 10 to 15 genera, and approximately 1,200 species. All lycophytes have simple leaves (**microphylls**) that contain a single strand of vascular tissue. Lycophytes include the club mosses, spike mosses, and quillworts.

Club Mosses

Club mosses (Lycopodiales) are a small group with 3 genera and approximately 380 to 400 species. The sporophyte is a low-growing terrestrial plant with simple branching stems (Figure 5.2). The plant spreads by a rhizome and has rudimentary roots. The small, scalelike leaves bear a resemblance to those of many conifers, hence the common name "ground pine" given to some club mosses. Leaves are opposite one another on the stem.

Spores are produced in sporangia on specialized leaves called **sporophylls,** which are often clustered at the ends of branches

Figure 5.2 *Lycopodium* is a typical club moss. Above, stiff club-moss (*Lycopodium annotinum*) is found in the Pocono Mountains, in Pennsylvania. Note the brown strobili at the ends of the branches.

into a structure called a **strobilus** or **cone**. Club moss gametophytes are either photosynthetic or obtain nutrition from organic material in the soil through symbiosis with mycorrhizal fungi.

Spike Mosses

The order Selaginellales has only 1 genus (*Selaginella*) and 700–750 species. The sporophyte is a branching rhizome similar to that of club mosses; whereas the leaves of club mosses grow opposite one another, spike mosses produce leaves in four rows: two rows of small leaves above two rows of larger leaves. These leaves have a scalelike structure on their base.

Although most spike mosses live in the tropics, some live in dry environments. These so-called resurrection plants become dormant when water is scarce, taking on a lifeless appearance, only to spring back to life when rehydrated.

Quillworts

Isoetales is the smallest lycophyte order, with only a single genus (*Isoetes*) and 125 species. Quillworts are relatively small aquatic plants that live underwater for part or all of their life. The sporophyte is a perennial **corm** that can be somewhat woody. Tufts of long, quill-like leaves, most of which are capable of producing sporangia, grow from the top of the corm.

Monilophytes

Monilophytes are a large diverse clade containing more than 12,000 species. All monilophytes produce **megaphylls**, leaves that have multiple strands of vascular tissue. There are six major groups within the monilophytes: horsetails, whisk ferns, ophioglossoid ferns, marattioid ferns, water ferns, and true ferns.

Whisk Ferns

Whisk ferns are absent from the fossil record. Psilotales is a very small order, with only 2 genera and 15–17 species. The genus *Psilotum* grows in tropical and subtropical areas. *Psilotum* has a simple body with small, scalelike leaves and a stem that branches off into two smaller stems. Plants have no roots; they spread by an underground rhizome covered with rhizoids. Both sporo-

Diminutive Descendents of Giants

Compared to other species of seedless vascular plants, quillworts are quite small, often being mistaken for clumps of grass—something that would never have happened to their lycophyte ancestors. Recent analyses have shown that the small quillworts of today are the living descendents of the giant lycophyte trees that once dominated swamp forests more than 350 million years ago. Why only small plants from the different seedless vascular plant lineages survived to the present is a mystery that botanists are still trying to solve.

phytes and gametophytes form relationships with mycorrhiozal fungi. *Psilotum* has distinctive sporangia that are borne in clusters of three on the stem.

Horsetails

Equisetales was once a diverse group, but it now contains only a single genus (*Equisetum*) with 15–25 species. The horsetails are **perennials** that grow in moist areas and along waterways. Although modern horsetails may reach heights of about 1 meter, ancient horsetails known as **calamites** grew to tree size.

Horsetails have a unique appearance. Branching rhizomes send up aerial stems, whose silica deposits give them a rough texture. The ridged stem is essentially a hollow tube of separate segments. At the joints of the segments there are rings of small, scaly leaves. Oval sporangia at the tips of the stems produce the spores.

Ophioglossoid Ferns

Ophioglossales, also called adder's tongue ferns, are a small group with 80–90 species. The leaves of these ferns are divided into two distinct regions: one that is photosynthetic and sterile, and another that is nonphotosynthetic and produces spores. Young leaves of Ophioglossoid ferns are not coiled as in other groups. They also produce upright stems, which differ from the horizontal stems of true ferns. The sporangia of ophioglossoid ferns develop from a group of cells, unlike the sporangia of true ferns that develop from a single cell.

Marattoid Ferns

Marattiales is also a small group, with only 6 genera and about 200 species. Marattoid ferns, like Ophioglossoids, produce upright stems. Their leaves are large, and when young, are coiled into a structure called a **fiddlehead** or **crozier** that uncoils as it opens (Figure 5.3). These ferns also produce a sporangium that develops from multiple cells. These ferns were common in swamps and are well represented in the fossil record.

Figure 5.3 The fiddlehead is an identifying feature of marattoid and true ferns. Above, a fiddlehead is seen unraveling.

Water Ferns

Water ferns grow in freshwater ponds and lakes. There are two different orders of water ferns, Marsileales and Salviniales. Both are small groups, with approximately 75 species in Marsileales and only 16 species in Salviniales. *Marsilea* plants grow in shallow water or mud and produce fronds with long **petioles** and leaves that look like a four-leaf clover. Ferns in Salviniales produce two types of fronds; one is flattened and floats on water, and the other is feathery and hangs below the surface of the water. The floating fronds contain colonies of cyanobacteria that fix nitrogen for the fern and, consequently, increase the nitrogen in the water where they grow. Despite this, water ferns can be pests in some areas. Both groups of water ferns encase their spores in a hard structure that can lie dormant in the mud for many years.

True Ferns

Filicales is the largest, most common group of seedless vascular plants. There are more than 11,000 species in 320 genera divided among 30–35 families. As mentioned previously, true ferns have several features that separate them from other fern groups. Their stems typically grow horizontally as rhizomes, and their young fronds form fiddleheads. True ferns are homosporous. Their sporangia develop from single cells on the undersurface

Noxious Weedy Ferns

Many ferns are attractive and desired plants. Others, however, are not. Two water ferns, *Salvinia molesta* (kariba-weed) and *Azolla pinnata* (feathered mosquito fern) are aggressive, noxious weeds that can clog waterways and cause flooding. Bracken fern (*Pteridium*) can invade pastures and choke out native pasture plants. This is particularly problematic because bracken fern fronds contain chemicals that are toxic to cattle that eat them.

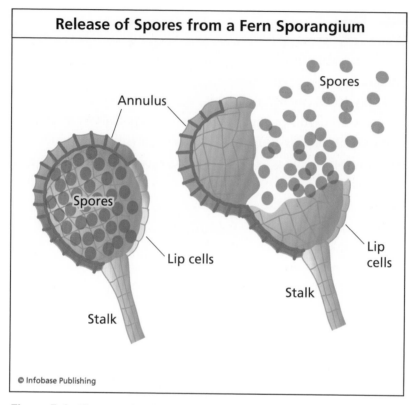

Release of Spores from a Fern Sporangium

Annulus

Spores

Spores

Lip cells

Lip cells

Stalk

Stalk

© Infobase Publishing

Figure 5.4   The annulus is a thickened area on the fern sporangium. Above, spores are released from the sporangium.

of fronds. Sporangia are clustered into structures called **sori** (singular: sorus) that are often covered by a flap of tissue called an **indusium**. Each sporangium has a thickened layer of cells called an **annulus** along one side of the sporangium (Figure 5.4). Tension generated by the annulus causes the mature sporangium to open and fling spores away from the parent plant.

True ferns grow in many temperate ecosystems, but they are particularly common in tropical areas. Ferns display a diversity of growth forms, including vines, small herbs, and even aquatic herbs. Large tree ferns grow in many tropical forests (Figure 5.5). Smaller ferns are common in the middle layers (known as

Figure 5.5 Tree ferns are the largest living, seedless vascular plants. The above image shows tree ferns in Malaysia.

the understory) of forests worldwide. Some fern species do not produce a sporophyte stage at all and exist only as gametophytes. Thus, these species only reproduce asexually.

ORIGINS AND RELATIONSHIPS OF SEEDLESS VASCULAR PLANTS

The oldest vascular plant fossils are of *Cooksonia* and date back approximately 410 million years (Table 4.2). These were small, simple, leafless plants that spread by branching rhizomes. They spread by a central stem that grew along the ground and produced upright branches with sporangia at their tips. They lacked roots, and appear to have used symbiotic relationships between their rhizoids and fungi to improve nutrient uptake from the soil. Woody growth evolved approximately 380 million years ago. Production of wood evolved independently in several different lineages.

The early vascular plant lineages Rhinophyta, Zosterophyllophyta, and Trimerophyta dominated the landscape 425 to 370 million years ago, but all three lineages were extinct by about 360 million years ago. During this time, many species became extinct due to the drying of the environment that occurred during that time. Only the ferns, herbaceous lycophytes, and horsetail lineages survived.

Seedless vascular plants can be divided into two distinct monophyletic clades (lycophytes and monilophytes) that are well supported by fossil, genetic, and structural data. The split between the lycophyte clade and other vascular plants probably occurred more than 400 million years ago. The split between the monilophyte clade and the seed plants occurred 20–30 million years later.

VALUE OF SEEDLESS VASCULAR PLANTS

Seedless vascular plants are components of many terrestrial ecosystems. Living seedless vascular plants are of limited economic value beyond their usefulness in landscaping and as houseplants.

Fiddleheads from some species are edible, and club moss spores were once used as an explosive in fireworks and in the development of photography. Long-dead seedless vascular plants that became coal, however, are of immense value as an effective fuel source.

SUMMARY

Seedless vascular plants evolved more than 400 million years ago. The evolution of vascular tissue enabled plants to become larger because of increased structural support and more efficient transport of materials around the plant body. Initially quite small, seedless vascular plants diversified and evolved many massive species that were once the giants of the ancient swamps. Presently, there are two major lineages of seedless vascular plants, the lycophytes and the monilophytes.

6 Nonflowering Seed Plants
The Gymnosperms

The pine tree lives a thousand years.
The morning glory flower lives a single day.
Yet both fulfill their destiny.
—Chinese proverb

Nonflowering Seed Plants
The Gymnosperms

The Namib Desert, one of the driest places on Earth, is home to one of the strangest plants on Earth. *Welwitschia mirabilis* survives by producing long taproots capable of accessing water from deep in the soil (Figure 6.1). The plants also collect water from cool fogs that blow in from the ocean. Droplets of water that condense on the plant's leaves flow down them to water the soil around the plant.

Unlike other members of the plant kingdom, *Welwitschia* produces only two leaves during its entire lifetime, which may last as long as 2,000 years. The leaves emerge from the germinating seed as the plant grows. Harsh desert winds shred the leaves, which continue to grow from their base. The leaves and reproductive structures known as strobili, or cones, grow from the edge of the woody, bowl-shaped stem.

Welwitschia belongs to a group of plants called the gymnosperms. The name *gymnosperm* comes from the Greek words *gymnos* (naked) and *sperma* (seed). These seeds are considered

Figure 6.1 The desert plant, *Welwitschia mirabilis,* is photographed in the Namib Desert in Namibia. An adult *Welwitschia* consists of two leaves, which are torn into strips when the leaves are whipped by strong winds.

Figure 6.2 Conifers are cone-bearing seed plants that display a variety of shapes and sizes. Above, conifers surround a lake in northern Wisconsin.

to be naked because they are not covered by sporophyte fruit tissues, as angiosperm seeds are. Scottish botanist Robert Brown (1773–1858) was the first to classify gymnosperms as a group separate from the other seed plants, based on these naked seeds.

There are approximately 840 species of gymnosperms represented by 4 divisions, the conifers (Coniferophyta; see Figure
6.2), cycads (Cycadophyta), ginkgos (Ginkgophyta), and gnetophytes (Gnetophyta). Some are restricted to warm tropical areas
while others form extensive forests in colder, higher latitudes and
areas of high elevation. Some gymnosperms are grown commercially for different uses.

FEATURES OF GYMNOSPERMS

Seeds and pollen are two features that have enhanced the reproductive success of gymnosperms, allowing them to evolve and
diversify. Seeds are an important evolutionary innovation
because they protect the dormant plant embryos until environmental signals cue **germination**. Seeds also contain stored
carbohydrates and nutrients that will feed the seedling as it
begins to grow.

Pollen is another critical trait to evolve in this group. Gymnosperms were the first plants to produce pollen as a way to
carry sperm to the egg for fertilization. Unlike bryophytes and
seedless vascular plants that release their sperm into the environment, the sperm cells of gymnosperms are encased in pollen
grains that are transported, typically by wind, to egg-contain-
ing **ovules** on female cones. When the pollen grain germinates,
a **pollen tube** grows into the ovule, and the sperm is released to
fertilize the egg.

Seeds and pollen are produced in cones (Figure 6.3). Similar
to the strobili in the seedless vascular plants, a gymnosperm
cone is a group of sporophylls attached to a short central axis.
Although sporophylls in most gymnosperms evolved as a modi-
fication of a leaf, the female conifer cone (pine cone) is actually
not a modified leaf, but rather a modified stem axis called a **cone
scale.**

Gymnosperms are vascular plants that produce xylem and
phloem. The gymnosperm stem and roots have a meristem called

Figure 6.3 An assortment of female pine cones are displayed above, including a pine cone with open scales *(a)*, a spruce pine cone *(b)*, an eastern white pine cone *(c)*, and a previous year seed cone beside clusters of male cones *(d)*.

the **vascular cambium**, which produces wood-forming xylem and bark-forming phloem. All gymnosperms are woody perennial plants. A majority of them grow as trees or shrubs, and many of them achieve awe-inspiring size and age. A vast majority of gymnosperms are **evergreen**, retaining living leaves on the plant throughout the year. A few, however, such as larch (*Larix*) and bald cypress (*Taxodium*) are **deciduous**, shedding their leaves once a year.

GYMNOSPERM LIFE CYCLE

The gymnosperm life cycle differs from those of bryophytes and ferns in its dramatic reduction of the gametophyte stage and increase in the sporophyte stage. The sporophyte is clearly the dominant phase. At no point in the gymnosperm life cycle is there a free-living, independent gametophyte.

The life cycle of pines (*Pinus)* serves as a good example of a typical gymnosperm life cycle. The mature pine tree is the adult sporophyte. Female cones are produced on the ends of branches. Each female cone is composed of numerous cone scales with female sporangia on them. In each female sporangium, a spore

Gymnosperms of Unusual Size and Age

Some of the oldest and largest organisms on Earth are gymnosperms. Several bristlecone pines (*Pinus longaeva*) in the White Mountains of eastern California are more than 4,700 years old. These plants, whose seeds germinated before the pyramids of Ancient Egypt were built, are the oldest known living organisms. A coast redwood (*Sequoia sempervirens*) in the Sierra Nevada Mountains of coastal California is the tallest tree on Earth, standing more than 365 feet (112 meters) tall and calculated to be more than 3,200 years old. The giant sequoia, (*Sequoiadendron giganteum*), which also grows in California, is another tall gymnosperm species. The tallest sequoias are more than 276 feet (84 meters) tall and 3,500 years old.

mother cell undergoes meiosis to form four spores, but only one survives. The single surviving spore undergoes numerous divisions by mitosis to form a multicellular female gametophyte surrounded by a thin layer of tissue called the **integuments**. The small opening in the integument is called the **micropyle**. The female gametophyte and integuments form a structure called the ovule. Inside the ovule are multiple archegonia, each containing an egg.

In the small, papery male cones (Figure 6.3d), spore mother cells undergo mitosis to form spores. Each spore develops into a pollen grain that contains a **tube cell** and a **generative cell**. Pine and other gymnosperm pollen grains often form air sacs that help carry them on the wind to the female cones. **Pollination** occurs when the pollen grain is either blown directly into the micropyle or trapped on a droplet of water secreted from the micropyle and drawn into it. The pollen grain makes contact with the female gametophyte, germinates, and slowly grows a pollen tube into the female gametophyte tissue. When the pollen tube reaches the archegonium (a process that can take up to a year), it releases the sperm cell to fertilize the egg. The embryo produced in the female gametophyte tissue begins to form a seed. As the seed develops during the course of another year, the female cone matures, forming the familiar woody pinecone. Once the seeds are mature, the cone scales open to release them, and the cycle begins again when the seed germinates to form a new seedling.

MAJOR GYMNOSPERM GROUPS

Gymnosperms are a relatively small group with approximately 800 living species (Table 6.1). Although there are few species, they are quite common and often maintain very large populations.

Conifers

With approximately 60 genera and more than 600 species, Coniferophyta is the largest, most well-known gymnosperm group. The largest family of conifers is the Pinaceae, which contains the pines (*Pinus*), spruces (*Picea*), hemlocks (*Tsuga*), and firs (*Abies*). Conifers typically produce evergreen, needlelike leaves, but the larches (*Larix*) produce deciduous leaves. Conifer leaves have many traits, such as a thick cuticle, sunken stomata, and modifications of the xylem, that help them survive in very cold and very dry environments.

There are approximately 100 species of *Pinus*. They are native to the Northern Hemisphere, where they grow in a diverse range of habitats from hot semideserts to cold mountain forests. Pines produce leaves in bundles called **fascicles**. The number of needles per fascicle can be an important trait for differentiating species. Male and female cones are produced on the same plant. Male cones are small, papery, and produce a large amount of pollen. The seed-producing female cones are woody and range in size from several millimeters to several centimeters in length. Female cones often have unique features that can be used to identify different species.

Another important conifer family is Cupressaceae. Junipers (*Juniperus*), cedars (*Cupressus*), redwoods (*Sequoia*), and other

Table 6.1   Major Groups of Living Gymnosperms

Group	Common name	Estimated species
Cycadophyta	cycads	130–150
Ginkgophyta	ginkgos	1
Coniferophyta	conifers	600–650
Gnetophyta	gnetophytes	70–80

members of this family are highly valued for their wood, which is used extensively in the construction industry for shingles and siding. Other species are commonly used as ornamentals in landscaping. Some species, particularly in *Juniperus,* produce fragrant wood that is used in making chests and closets. Cupressaceae is an old family whose lineage dates back 248 to 206 million years. *Sequoiadendron giganteum* (giant sequoia) and *Sequoia sempervirens* (coast redwood) of California are some of the largest trees on Earth.

The conifers listed previously are predominantly found in the Northern Hemisphere. The families Podocarpaceae and Araucariaceae, on the other hand, are common in the Southern Hemisphere. The seeds of these families are often associated with a fleshy structure that attracts birds. The monkey puzzle tree (*Araucaria*) produces some of the largest trees in southern tropical forests. The Norfolk Island pine (*Araucaria heterophylla*) is a common ornamental species.

Cycads

Cycadophyta is a very old lineage of tropical and subtropical plants that evolved 290 to 248 million years ago. They were an extremely diverse group, but have since dwindled to approximately 130–150 species in 11 genera.

Cycads produce large leaves that are often mistaken for palms. Like all gymnosperms, cycads reproduce using cones. Unlike conifers, however, cycads produce male and female cones on different plants. They are pollinated by insects, predominantly beetles, which transfer pollen from male to female cones while feeding on pollen and other cone parts. After pollination, the pollen tube grows into the ovule and releases flagellated sperm that swim to the egg for fertilization.

Ginkgos

Ginkgophyta evolved 260 million years ago. Although diverse and widespread in the fossil record, only one species, *Ginkgo*

Figure 6.4 Shown above are *Ginkgo biloba* leaves and female cones. *Ginkgo biloba* is an ancient gymnosperm species, whose extract is commonly used as a memory enhancer.

biloba (ginkgo or maidenhair tree), still survives (Figure 6.4). Comparisons of living plants to fossils show that this species has changed very little during the past 100 million years.

Ginkgos are attractive trees. Their deciduous fan-shaped leaves turn a brilliant yellow in autumn. Male and female cones are borne on separate trees. After pollination, the pollen grains form highly branched pollen tubes that grow throughout the ovule tissue. Once the pollen tube contacts the archegonium, it

Living Fossils

Paleobotanists recognized species of ginkgos and dawn redwoods from fossil records in which they were very abundant. In 1691, living ginkgo trees were "discovered" to Westerners by Engelbert Kaempfer, who found them growing in the temple grounds of Buddhist monks in China and Japan. In 1941, an unknown tree species was seen in a forest in China. In 1948, these trees were identified as dawn redwoods, previously believed to be extinct. Finally, in 1994, 39 trees of a gymnosperm species (*Wollemia nobilis*, Araucariaceae) thought to be extinct were found growing in a remote forest in Australia.

bursts and releases the flagellated sperm that swim to the egg for fertilization. The fleshy female ginkgo cones are infamous for their stench, resembling a combination of old cheese and feet. Hence, most ornamental plantings consist of male trees.

Gnetophytes

Gnetophytes are quite distinct in appearance from one another and other gymnosperms. Structural and genetic data strongly support Gnetophyta as a monophyletic group. This lineage contains three genera: *Welwitschia, Gnetum,* and *Ephedra.*

Welwitschia mirabilis, described at the beginning of this chapter, is the lone member of this genus. Plants in the genus *Gnetum* are woody tropical vines whose leaves resemble the broad leaves of angiosperms. There are approximately 30 *Gnetum* species.

The remaining genus, *Ephedra* (Mormon tea) has approximately 40 species, most of which grow in the arid regions of the western United States and Mexico (Figure 6.5). They are typically shrubby plants with jointed photosynthetic stems and small, scalelike leaves. Ephedras produce the chemical ephedrine, which acts as a stimulant and can be used to treat allergies and respiratory disorders.

Figure 6.5 Ephedra, commonly referred to as Mormon tea, are shrubby plants that grow in arid environments, such as Arches National Park in Utah *(above)*. Mormon tea was used by Native Americans to make a stimulating drink.

Gnetophytes are of particular importance to botanists because they share interesting similarities with flowering plants. For example, male cones produce pollen in structures that are similar to anthers found in flowers. Likewise, modified leaves called **bracts** are often associated with male and female gnetophyte cones, much the same way that modified leaves are associated with flowers. Most important, however, the reproductive process of gnetophytes is very similar to that of angiosperms.

In all other gymnosperms, the pollen grain produces two sperm cells, only one of which fertilizes the egg. In gnetophytes, one sperm fertilizes the egg and the other fuses with a non-egg cell in the female gametophyte. This second fertilization forms a short-lived embryo that eventually degenerates, but it may provide some nourishment to the other developing embryo. This process of two fertilization events is similar to the one in angiosperms, and may suggest a common ancestor between gnetophytes and angiosperms. Some structural data suggest, however, that gnetophytes are more closely related to pines and are not ancestors to the angiosperms.

ORIGINS AND RELATIONSHIPS OF GYMNOSPERMS

Fossil evidence indicates that seed plants first evolved approximately 365 to 380 million years ago. The oldest known gymnosperm fossils are of a plant in the genus *Archaeopteris*. It is a member of a group referred to as the **progymnosperms**, which means "before the gymnosperms." *Archaeopteris* was a treelike, woody plant. It had leaves similar to fern fronds and produced seeds in sporangia attached to the branches. Other progymnosperm fossils have features indicating that they evolved from seedless vascular plants. Seed plants evolved from these progymnosperm ancestors.

Several diverse lineages of seed-producing plants evolved during the 65 million years that followed. One group called **seed ferns** had fernlike leaves that produced seeds at their tips.

Another important group, the cordaites (Cordaitales), contained trees and shrubs that resembled modern conifers. Seed ferns and cordaites became extinct 290 to 248 million years ago, but they, along with lycophytes and calamites, were dominant plants before that time.

Cycadeoids (Bennettitales) are another extinct gymnosperm lineage. Cycadeoids are interesting because they are the only gymnosperm lineage to produce cones containing both male and female structures. Cycadioids, cycads, and ginkgos evolved 248 to 206 million years ago and diversified 206 to 144 million years ago. Cycadeoids became extinct approximately 65 million years ago, but the two other lineages exist still today. Although they have lost some of their original diversity, this highly successful group of plants has existed for more than 145 million years.

Because some of the earliest gymnosperm lineages are known only from fossils, there is uncertainty about the relationships among them. Several different phylogenies have been proposed for living gymnosperms, each with its own strengths and weaknesses, depending upon the data used to construct the phylogeny. As described earlier, some botanists have concluded that gnetophytes are the gymnosperm lineage that shared the most recent common ancestor to angiosperms because of their similar fertilization process (called the *anthophyte hypothesis*). Although there are compelling reasons that support this conclusion, current structural and genetic data suggests that angiosperms diverged from the gymnosperms lineage very early, and that gymnosperms continued to diversify (called the *gne-pine hypothesis*). Cycads and ginkgos are closely related basal lineages in the gymnosperms. The presence of flagellated sperm in these two groups also indicates that these are older gymnosperm lineages. Nonpine conifers comprise one remaining lineage, and Pinaceae conifers and gnetophytes evolved from a common ancestor.

Every Tree Is a Hypothesis

A fundamental component of the scientific method is the developing and testing of hypotheses. In systematics, the hypothesis is the phylogeny developed by a systematist. Because it is impossible to actually observe the entire evolutionary history of a group of plants, systematists develop a phylogeny based on their interpretations and conclusions from the data they collected. Other systematists can then collect and analyze additional data to test the relationship proposed in the original phylogeny or propose new relationships.

VALUE OF GYMNOSPERMS

Gymnosperms are ecologically important plants that are dominant species in many ecosystems worldwide. In addition to providing habitat and food for other organisms, they also shape their environment by affecting soil, water availability, and temperature.

Gymnosperms have significant economic value. A wide variety of species are used in landscaping. Other species are grown for fuel, lumber, and paper pulp production. Pines in particular are tapped for resins, which can be converted into a variety of commercial products including turpentine and rosins. In the past, pine resins called naval stores were used as a waterproofing agent for wooden ships. A few gymnosperms, such as piñon pines and junipers, are even grown for food and flavorings.

SUMMARY

Gymnosperms are seed-producing vascular plants. Seeds and pollen are produced in cones composed of sporophylls or cone scales. Gymnosperms arose from seed-producing ancestors that evolved more than 365 million years ago. Gymnosperm lineages, such as the seed ferns, cordaites, and cycadeoids, were large, diverse groups that have since become extinct. Conifers, cycads, ginkgos, and gnetophytes are the four living gymnosperm lineages.

7 Flowering Plants
The Angiosperms

The rapid development . . . of all the higher plants within recent geological times is an abominable mystery.

—Charles Darwin (1809–1882)
English naturalist

Flowering Plants
The Angiosperms

The passage on the previous page is from a letter written by Charles Darwin to his colleague, the eminent botanist Sir Joseph Hooker. The "abominable mystery" Darwin referred to was the sudden and dramatic appearance and diversification of angiosperms in the fossil record. Fossils of other major groups of plants show a pattern of lineages appearing gradually and diversifying slowly over the course of hundreds of millions of years. Angiosperms made their first appearance in the fossil record approximately 135 million years ago. Angiosperms then underwent a rapid proliferation and diversification 70–100 million years ago. By 60 million years ago, a majority of the modern angiosperm families that exist today were well established. No other group of plants has shown such rapid evolution, diversification, and dominance of the landscape in the history of Earth.

The success of the angiosperms was made possible largely through the unique features that resulted from the evolution of flowers and fruits. A survey of the more than 250,000 species of angiosperms reveals an astonishing array of flower and fruit structures and modifications. Another evolutionary advantage of angiosperms is their diversity of growth forms and vegetative traits. Angiosperms range from minute aquatic plants,

Say It With Flowers

"O, my love's like a red, red rose...." The Scottish poet Robert Burns, and millions of other romantics past and present, have associate red roses with true love. Throughout history, flowers and fruits have been associated with emotions and other things in what is called "The Language of Flowers." Acorns are a symbol of immortality. Foxgloves represent insincerity. Yellow poppies and corn symbolize success and riches. Several nations are also associated with flowers: peonies with China, lilies with France, and thistles with Scotland.

such as watermeal (*Wolffia*), to a giant quaking aspen (*Populus tremuloides*) clone that covers 106 acres and is the largest organism on Earth. Their ability to adapt to their environment has allowed angiosperms to have an amazing impact on the ecology and evolution of life on Earth. Angiosperms are an important resource to humans, as they have both practical use and cultural significance.

THE FLOWER

The word *angiosperm* comes from the Greek words *angeion* ("vessel") and *sperma* ("seed"), and refers to the fact that

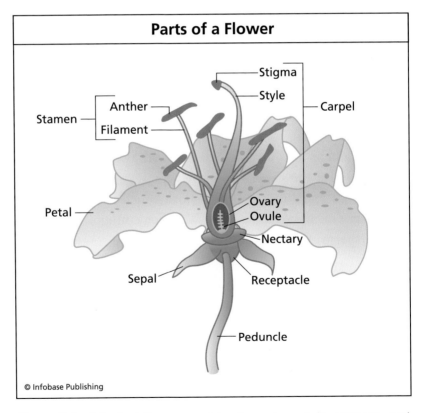

Figure 7.1 A typical flower consists of sepals, petals, stamens, and one or more carpels.

angiosperms produce seeds from ovules that are completely enclosed in sporophyte tissue. This is in contrast to gymnosperms, which produce seeds from a bare ovule. Specifically, the angiosperm vessel is the wall of the **ovary** that encloses the seed. The presence of flowers that produce these seed-enclosed fruits is a defining characteristic of angiosperms.

Flowers consist of four parts (**sepals, petals, stamens,** and **carpels**) that are arranged into four groups called **whorls** (Figure 7.1). All four whorls attach to a structure at the base of the flower called the **receptacle**. The outermost whorl, or **calyx**, is composed of green, leaflike sepals, which cover and protect the developing flower **bud**. The petals of the flower make up the next whorl, which is called the **corolla**. Petals may resemble leaves, but unlike leaves, they are often brightly colored, attracting **pollinators** for sexual reproduction. Sepals and petals that cannot be distinguished from one another, as is the case with tulips and cacti, are called **tepals**. The calyx and corolla are called the **sterile whorls** because they do not produce gametes.

The stamens and carpels make up the two fertile whorls, which produce the gametes required for sexual reproduction. The stamens make up the third whorl, which is known as the **androecium** (meaning "male house"). Pollen containing the male **gamete** (sperm) is produced in the **anther** of the stamen, which is supported on a stalk called the **filament**. The innermost whorl of a flower is the **gynoecium** (meaning "female house"), which is composed of the carpels. Carpels are made of a **stigma, style,** and ovary. The carpel is also called the **pistil**. A gynoecium may consist of one or more carpels. A gynoecium composed of two or more carpels fused together is a **compound pistil**. The ovary of the gynoecium eventually matures to form a fruit containing one or more seeds.

Flowers are grouped together on a plant to form an **inflorescence** (Figure 7.2). An inflorescence can be composed of a single

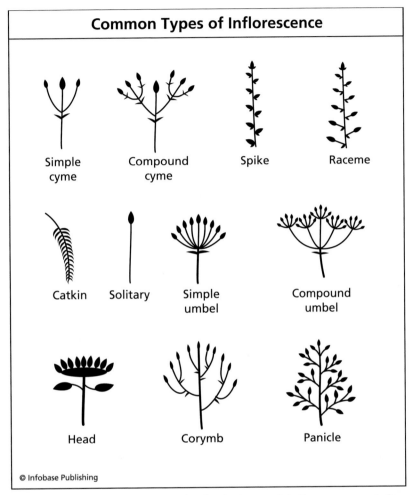

Common Types of Inflorescence

Simple cyme   Compound cyme   Spike   Raceme

Catkin   Solitary   Simple umbel   Compound umbel

Head   Corymb   Panicle

© Infobase Publishing

Figure 7.2 Shown above are the basic types of inflorescences. Inflorescence refers to the way in which individual flowers are arranged on the stem.

flower (solitary inflorescence), or any number of different multi-flowered inflorescences. The stalk that supports an inflorescence, whether multi-flowered or solitary, is called the **peduncle.** The stalk that supports an individual flower within a multi-flowered inflorescence is called the **pedicel.**

ANGIOSPERM LIFE CYCLE

The life cycle of a lily serves as a good example of the angiosperm alternation of generations. It begins with the germination of a seed, which is the sporophyte generation. The seedling develops into a mature plant that will eventually form flowers. Most flowers contain both stamens and carpels. In some angiosperms, stamens and carpels are produced in separate flowers on the same or different plants.

Within the anther on the stamen, spore mother cells undergo meiosis to form haploid spores. Each spore develops into a pollen grain containing a tube cell and a generative cell. When the pollen grain germinates on the stigma, the tube cell controls the growth of the pollen tube. This tube grows through the style toward the ovary, carrying with it the generative cell. The generative cell divides to form two sperm cells. The growing pollen tube is the male gametophyte.

Within the carpel, spore mother cells undergo meiosis to form four haploid spores. Three of the haploid spores degenerate while one divides three more times to produce the female gametophyte. The angiosperm female gametophyte is composed of seven cells. The large **central cell** contains the two haploid **polar nuclei**. Three cells are the **antipodals** and two cells called the **synergids** are next to the egg. These seven cells and eight nuclei make up the mature female gametophyte and, with the integument layers around it, form the ovule. Ovules are enclosed and attached to the sporophyte tissues of the ovary. Thus, the female gametophyte is physically attached to, and nutritionally dependent upon, the mature sporophyte.

A key feature of the angiosperm life cycle is the reduction of the gametophyte stage, particularly the female gametophyte, and elaboration of the sporophyte. This is exactly the opposite of what occurs in bryophytes, where the sporophyte is attached to, and nutritionally dependent upon a much larger gametophyte.

Pollination by Deception

Flowers use different combinations of color, scent, and shape to attract pollinators. Some of the most specialized flower pollinator relationships involve orchids, whose flowers mimic the color, shape, and even scent of females in the bee and wasp species that pollinate them. Male bees and wasps, lured in by the false promise of a mate, inadvertently transfer pollen from one flower to another. Some arums (plants of the family Araceae) use a similar strategy, mimicking the stench of rotting meat. Flies, seeking a place to deposit their eggs, pollinate the flowers.

DOUBLE FERTILIZATION

When the pollen tube reaches the ovule, it releases the two sperm cells. One sperm cell fuses with the egg to form the zygote that will eventually become the next sporophyte. The other sperm cell fuses with the two polar nuclei to form the **endosperm**, which provides food for the developing zygote. This **double fertilization** is a unique feature of the angiosperm life cycle.

Following the formation of the zygote and endosperm, the ovule develops into the seed that contains the plant embryo and food reserves, which may be in the form of endosperm or sugars and starches stored in modified leaves called **cotyledons**. Ripe ovary tissues surrounding the seed form the fruit. Ultimately, the seed is dispersed from the plant through a variety of different mechanisms. Some fruits are fleshy and contain sugars, lipids, or other substances to entice animals to eat the fruit and internally transport the seed until it passes through the digestive tract. Others have hooks, barbs, or other devices that attach to the outside of animals to disperse the seed. Once dispersed, the seed will germinate and form the next sporophyte.

MAJOR ANGIOSPERM GROUPS

In 1703, English botanist John Ray divided angiosperms into **monocots** (plants whose seeds have one cotyledon) and **dicots**

Table 7.1   Major Groups of Living Angiosperms

Group	Common name	Estimated species
Amborellaceae		
Nymphaeaceae		
Autrobaileyales	basal angiosperms	170
Nymphaeaceae, Illiciaceae		
Magnoliopsida*	magnoliids	9,000
Liliopsida	monocots	70,000–72,000
Magnoliopsida*	eudicots	175,000–180,000

* Magnoliopsida is the traditional name given to dicots. At present there is no ICBN name for the separate magnoliid or eudicot clades.

(plants whose seeds have two cotyledons). Botanists have traditionally followed this classification and continued to divide angiosperms into the Magnoliopsida (dicots) and Liliopsida (monocots) based on a variety of traits, such as the arrangement of **vascular bundles** (Figure 7.3) and features of roots, leaves, and flowers.

Recent studies, however, have shown that the dicots are an artificial grouping of several different lineages. Some plants previously classified as dicots are actually ancient angiosperm lineages that are now collectively referred to as the **basal angiosperms** (Table 7.1). This group consists of several families, including the Amborellaceae, Nymphaeaceae, and Illiciaceae. The remaining angiosperms are now referred to as the **core**

angiosperms. This group includes three monophyletic lineages: magnoliids, eudicots, and monocots. The magnoliid clade contains several ancient angiosperm families, including the

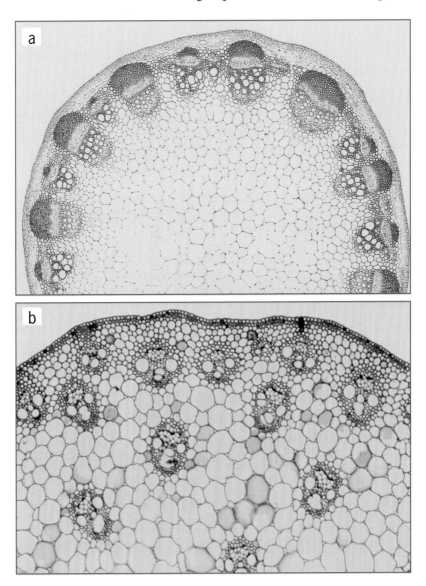

Figure 7.3 Cross-sections of monocot and dicot stems reveal the differences in vascular bundle arrangement. Vascular bundles are arranged in a ring in dicots *(a)*, but scattered in the monocot stem *(b)*.

magnolias (Magnoliaceae), laurels (Lauraceae), spicebushes (Calycanthaceae), and black peppers (Piperaceae). **Eudicots** (true dicots) make up a large, natural group that contains about 75% of all dicot species. Eudicots share a common feature that is not found in any other flowering plants: They have three openings in their pollen grain walls. All other plants (gymnosperms, basal angiosperms, magnoliids, and monocots) produce pollen grains that have only one opening in their walls. Beyond this difference, magnoliids and eudicots share many other features (Table 7.2).

Despite changes in our understanding of dicots, monocots are still recognized as a single lineage that descended from a common ancestor. Monocots account for approximately 22% of all angiosperms.

Table 7.2 **Differences Between the Three Core Angiosperm Groups**

Trait	Monocots	Magnoliids	Eudicots
Cotyledons	1	2	2
Vascular bundle arrangement in stem	scattered	ring	ring
Leaf vein pattern	parallel	pinnate	palmate, pinnate
Number of flower parts	three-merous	many	four-merous, five-merous
Openings in pollen grains	1	1	3

Significant differences exist between the floral structures of eudicots and monocots. In eudicot flowers, the parts of the different floral whorls are typically in multiples of four (**four-merous**) or five (**five-merous**). For example, members of the mustard family (Brassicaceae) produce flowers with four sepals, four petals, and four stamens. Species in the rose family (Rosaceae), such as roses, strawberries, and blackberries, have flowers with five sepals and five petals. Monocot flowers, such as irises, lilies, and tulips, are **three-merous**, with three sepals, three petals, three or six stamens, and a pistil with three carpels.

ECOLOGICAL AND ECONOMIC VALUE

Angiosperms are dominant components of terrestrial environments worldwide and, consequently, play a major role in the ecological dynamics that maintain a healthy planet. Angiosperms also include many important resources that humans depend on for a variety of purposes. Angiosperms provide food, ornamentals, fibers, and a wide array of chemicals.

Oaks (Fagaceae), hickories (Juglandaceae), and maples (Sapindaceae) are major components of forests in many temperate areas. Tropical forests, however, are often dominated by trees from a multitude of different families, including dipterocarps (Dipterocarpaceae), laurels (Lauraceae), and figs (Moraceae). Trees in these families support numerous other life forms in forest ecosystems by providing food and habitats. Many of these families also provide timber and other resources for humans.

The rose family (Rosaceae) is well known for its ornamental species, like roses (*Rosa* sp.) and cinquefoil (*Potentilla*), and also for its edible species (Figure 7.4), like apples (*Malus*), and pears (*Pyrus*), and peaches, plums, and cherries (all in the genus *Prunus*).

The bean family (Fabaceae) is one of the largest angiosperm families. Members of Fabaceae are also called **legumes**, which refers to the unique fruit produced by species in this family. The

Figure 7.4 The rose family includes many edible fruits. Pictured above are two ripened apples *(a)*, peaches ready to be picked *(b)*, ripe cherries *(c)*, and fresh pears *(d)*.

symbiotic relationships between legumes and **nitrogen-fixing** soil bacteria give legume seeds (beans and peas) a high protein content, making this an agriculturally important family. Species such as beans (*Phaseolus vulgaris*), soybeans (*Glycine max*), field peas (*Vicia faba*), and lentils (*Lens culinaris*) contribute to the diets of humans and wildlife worldwide.

The mint family (Lamiaceae) includes many culinary herbs and plants renowned for their perfumes. The flavors and scents of plants in the mint family come from glands on the leaves, which produce aromatic oils. Basil (*Ocimum basilicum*), lavender (*Lavandula*), mint (*Mentha*), oregano (*Origanum vulgare*), patchouli (*Pogostemon cablin*), sage (*Salvia officinalis*), rosemary (*Rosmarinus officinalis*), and thyme (*Thymus*) are valued herbs in Lamiaceae. Catnip (*Nepeta*), coleus (*Plectranthus*), and bee balm (*Monarda*) are popular ornamentals.

The nightshade family (Solanaceae) contains many poisonous plants. The poisons are often powerful **alkaloids,** which have a range of effects. Tobacco (*Nicotiana*), belladonna (*Atropa*), and jimsonweed (*Datura*) produce alkaloids with narcotic effects. Hot peppers (*Capsicum*) get their heat from the alkaloid capsaicin. Important edible plants, including tomato (*Solanum lycopersicum*) and potato (*Solanum tuberosum*), are also in this family.

The sunflower family (Asteraceae) is the largest of all eudicot families, with more than 25,000 species. It contains many commercially important species, such as the annual sunflower (*Helianthus annuus*), which is grown for its seeds, oil, and flowers. This family also contains commercially grown ornamentals, including dahlias (*Dahlia*), mums (*Chrysanthemum*), and marigolds (*Tagetes*), as well as aggressive, noxious weeds, such as dandelion (*Taraxacum officiale*) and Canada thistle (*Cirsium arvense*).

The grasses (Poaceae) account for approximately 17% of all angiosperms. This monocot family serves as the foundation of food webs and other processes in prairies, savannas, and tundra ecosystems. Grasses can be found in most other ecosystems as well. Nine of the top 20 crop species are grasses, with corn (*Zea mays*), wheat (*Triticum aestivum*), and rice (*Oryza sativa*) providing 50% of human caloric intake. Bamboos (*Bambusa*) are grasses that have a wide range of uses, including construction, textiles, and papermaking. Other species, such as cheat grass (*Bromus tectorum*) and Johnson grass (*Sorghum halapense*), are invasive species that have a detrimental effect on native vegetation.

Figure 7.5 The lilies, irises, and orchids produce bright, showy flowers. Photographed above are the white lily *(a)*, purple iris *(b)*, and the harlequin "dancing lady" orchid *(c)*, also known as a butterfly plant.

The irises (Iridaceae), orchids (Orchidaceae), and lilies (Liliaceae) are often called the "showy monocots" because of the bright, colorful petals in their flowers (Figure 7.5). The Iridaceae includes irises (*Iris*), gladiolus (*Gladiolus*), crocus (*Crocus*), and blue-eyed grass (*Sisyrynchium*). The anthers from *Crocus sativus* are the source of the spice saffron. Orchids are the largest angio-sperm family, with more than 30,000 species. There are terrestrial orchids as well as many **epiphytes,** plants that grow on the stems and branches of trees. Orchids are particularly diverse in the trop-ics, where they form close relationships with the organisms that pollinate their flowers. Liliaceae contains ornamentals including lilies (*Lilium*) and tulips (*Tulipa*). There are several edible lilies, including onion (*Allium cepa*), garlic (*Allium longicuspis*), and chives (*Allium schoenoprasum*).

ORIGINS AND RELATIONSHIPS OF ANGIOSPERMS

The oldest angiosperm fossils are flowers and pollen that are 130–140 million years old. The oldest complete fossil plant is in the genus *Archaefructus* that lived more than 125 million years ago. It was a simple plant whose flowers had no sepals or petals, but numerous stamens.

One of the first hypotheses about the origins of flowering plants was developed by Engler and Prantl, who suggested that Amentiferae, a group in their classification, was the most primi-tive angiosperm group. This conclusion was based on the fact that many of these trees are wind pollinated, just like the more primitive gymnosperms. Other researchers concluded that fami-lies such as Magnoliaceae (magnolia family) and Ranunculaceae (buttercup family) were the most primitive because their flow-ers contain many parts. Recent genetic analyses have shown, however, that neither of these families are representative of the oldest type of angiosperm. Magnoliaceae is an ancient family, but it is not the oldest, and Ranunculaceae has been shown to be a member of the eudicots.

Currently, many botanists agree that the families Amborel-laceae, Nymphaeaceae, Trimeniaceae, and the order Illiciales are the oldest living lineages of flowering plants. These groups produce small, simple flowers that have few parts and simple pollination mechanisms. Collectively, these ancient groups form the basal angiosperms. Fossil evidence also indicates that the first angiosperms had small, nondescript flowers with few floral parts. The species *Amborella trichopoda* (Amborellaceae) is a shrubby plant that grows in the cloud forests of New Caledonia. Many studies have indicated that it is likely the living descendent of the oldest angiosperm lineage on Earth. Fossil and genetic data indicate that monocots, magnoliids, and eudicots diverged as separate lineages approximately 100 million years ago.

ANGIOSPERM SUCCESS

Angiosperms have evolved and dominated the terrestrial environment like no other group of plants. The flower undoubtedly is a primary reason for this success. Flowers enabled plants to use different shapes, colors, and scents to attract pollinators. Flowers also frequently provide a food reward of pollen or nectar to pollinators in return for the service of visiting flowers and transferring pollen. The evolution and diversification of these insect and bird pollinators is linked to that of the angiosperms.

Like flowers, fruits provide a food reward to animals that eat a fruit and disperse the enclosed seeds. Fruit size, shape, and color have evolved in response to natural selection mediated by

Big Flowers

The brownish-orange flowers of the giant rafflesia (*Rafflesia arnoldi*) are the largest single flowers of any angiosperm. A typical *Rafflesia* flower can be up to 3 feet (1 m) across and weigh as much as 25 pounds (11 kg). The titan arum (*Amorphophallus titanium*) produces the largest inflorescence. Its flowers are very small, but they combine to form a special type of inflorescence called a **spadix.** Separate male and female flowers are attached to a central column that can be 3 to 6 feet (1 to 2 m) tall. A huge modified leaf surrounding the column gives the spadix the appearance of being a single large flower.

the animals that eat and disperse the seeds. Angiosperms have evolved an array of mechanisms to disperse their offspring.

Changes in the life cycle have also promoted angiosperm evolution. While the other seed plants, gymnosperms, can take as long as two years to progress from pollination to the release of the mature seed, angiosperms have accelerated the process. Pollination, fertilization, and maturation of the seed can occur within a matter of days or months. This acceleration may be due to the nutritional benefits of the endosperm for the developing embryo. Angiosperms are also the only seed plants to have evolved the annual life cycle in which a plant completes its life in one growing season. By shortening different aspects of their life cycle, angiosperms can reproduce more quickly, thereby speeding their response to natural selection and accelerating the evolution and spread of adaptive traits.

SUMMARY

Angiosperms are vascular plants that produce seeds from flowers and enclose the seeds in a fruit. Angiosperms are the most recently evolved group of plants, and they dominate the terrestrial environment. The flower and fruit are key traits that allowed angiosperms to interact with animal pollinators and seed dispersers and, consequently, undergo extensive diversification. There are three major groups of angiosperms: basal angiosperms, monocots, and eudicots. Many species in these groups are ecologically and economically important.

Glossary

Algae A general term for photosynthetic members of the kingdom Protista.

Alkaloid A type of chemical produced by certain plants that is made out of nitrogen and can be toxic when ingested.

Alternation of generations A life cycle that alternates between a multicellular haploid generation and a multicellular diploid generation.

Anatomy The study of an organism's structure.

Androecium The male parts of a flower.

Angiosperms Plants that produce flowers and whose seeds are contained within a fruit.

Animalia The animal kingdom.

Annulus A thickened band of cells on a fern sporangium, which is involved in the dispersal of spores.

Anther Floral structure in which pollen is produced.

Antheridia Male reproductive structures that produce and protect sperm.

Antipodals The three cells of the female gametophyte that are clustered at the end opposite the egg.

Arbuscules Specialized structures formed by mycorrhizal fungi inside a plant cell that enable the plant and fungus to exchange nutrients

Archaea The kingdom of prokaryotic unicellular organisms containing blue-green algae, methane-producing bacteria, and other primitive life forms.

Archegonia The female reproductive structures that produce and protect the egg.

Artificial classification system A system of grouping organisms that is based on arbitrary features rather true evolutionary relationships.

Asexual reproduction Formation of offspring that does not involve the joining of egg and sperm.

Ascus (plural: asci) A saclike cell in which ascomycetes form spores.

Ascocarp The fruiting body of an ascomycete.

Autotrophs Organisms capable of synthesizing its own energy.

Bacteria A group of unicellular, prokaryotic organisms.

Basal angiosperms A name given to the most primitive angiosperm lineages.

Basidium (plural: basidia) The club-shaped structure in which basiomycetes form spores.

Basidiocarp The fruiting body of a basidiomycete.

Binomial nomenclature The system developed by Linnaeus in which each species is given a name consisting of two words: the genus name and the species name.

Biodiversity The diversity of life on Earth.

Blades The flattened photosynthetic regions of a leaf or algae.

Bracts Modified leaflike structures that can be part of a flower or fruit.

Brown algae Group of multicellular algae usually found in marine environments.

Bryophytes Non-flowering plants that include mosses, liverworts, hornworts, and quillworts.

Bud Small embryonic floral or vegetative shoots.

Calamites Ancient group of horsetails.

Calyx The outermost floral whorl consisting of all of a flower's sepals.

Capsule A spore-containing structure of the moss sporophyte.

Carotenoids A class of yellow and orange pigments.

Carpels The ovule-encasing structures that make up the gynoecium.

Cellulose The main component of cell walls in plants.

Central cell The cell in the angiosperm female gametophyte that contains two haploid nuclei; it becomes endosperm after fusion with second sperm nucleus.

Chitin A substance that forms the cell walls of certain fungi.

Chlorophyll A green pigment responsible for capturing the light used in photosynthesis. There are two types of chlorophyll, *a* and *b,* which have slight differences in the wavelengths of light they absorb.

Chloroplasts Structures within plant cells that contain the enzymes and pigments necessary for photosynthesis.

Chromosomes The organized structures of DNA in a cell.

Clade A group consisting of an ancestor and all of its descendants.

Glossary

Cladistics The system of grouping organisms based on an analysis of their primitive and advanced traits.

Classification The process of organizing groups into a hierarchical system.

Common name The unofficial name given to a plant by those who live near it.

Compound pistil A gynoecium composed of two or more carpels fused together.

Cone A reproductive structure that produces either pollen or seeds, typically found in gymnosperms and other groups of nonflowering plants.

Cone scale An individual unit of the gymnosperm strobilus.

Core angiosperms Groupings that include three monophyletic lineages: magnoliids, eudicots, and monocots.

Corm A dry underground structure found in perennial plants such as gladiolus.

Corolla The (usually) conspicuously-colored flower whorl consisting of all of the flower's petals.

Cotyledon A leaf that provides nourishment to the plant embryo during germination.

Crozier The coiled leaf of a fern.

Crustose A lichen growth form that resembles a crust on the substrate upon which it grows.

Cuticle A waxy, protective layer on the outer surfaces of leaves.

Deciduous A plant that loses its leaves during autumn or the dry season.

Dendrograms Treelike diagrams that illustrate relationships.

Dichotomous key A list of paired, contrasting statements used to identify an unknown organism by the process of elimination.

Dicots A general term given to angiosperms whose seeds produce two cotyledons, now divided into two groups, the basal angiosperms and the eudicots.

Dikaryotic Cells in fungi that have two haploid nuclei.

Diploid Having two full sets of chromosomes in each cell, characteristic of the sporophyte generation.

Diversity A measure of the number and relative proportion of different species in a community.

Domain The taxonomic level above kingdom.

Double fertilization A two-fertilization event, characteristic of angiosperms, wherein one sperm fuses with the egg to form the zygote and a second sperm fuses with the polar nuclei to form the endosperm.

Endosperm The tissues formed during a double fertilization event that nourish the growing angiosperm embryo.

Epiphytes Nonparasitic plants that grow on trees.

Ethnobotanists Scientists who study the ways in which indigenous peoples use plants, particularly for medicinal purposes.

Eudicots The largest group of angiosperms; they have two cotyledons and three openings in their pollen grains.

Eukaryotic Organisms whose cells contain nuclei and membrane-bound organelles.

Evergreen A tree whose leaves can be shed at any time of year, but never all at once.

Evolution A change in genetically-based characteristics of a species throughout time.

Fascicles Bundles of pine needles.

Fertile whorls The whorls of a flower that produce gametes.

Fertilization The joining of egg and sperm in sexual reproduction.

Fiddlehead *See* Crozier.

Field guide A book that contains simple descriptions and illustrations of plants in an area.

Filament The stalk of a stamen.

Five-merous A flower that contains parts in multiples of five.

Flagella Tail-like structures that propel many eukaryotic cells, including sperm.

Flavonoids A class of pigments found only in plants and a few algae.

Float The inflated part of a kelp body.

Glossary

Flora A guide with descriptions, illustrations, distributions maps, and keys that allow the user to identify plants of a particular geographical area.

Flowers The reproductive structures of angiosperms composed of sepals, petals, stamens, and carpels.

Foliose A lichen growth that resembles leaves.

Food webs An interlocking system of producers and consumers in an ecosystem.

Four-merous A flower that contains parts in multiples of four.

Fronds Fern leaves.

Fruit The mature seed-containing ovary (or group of ovaries) of angiosperms.

Fruticose A lichen growth form that is highly branched and erect.

Fungi The kingdom of organisms that have cell walls and obtain their food through absorption.

Gamete A haploid reproductive cell, either sperm or egg.

Gametophyte The haploid, gamete-producing stage in the alternation of generations.

Gemmae (singular: gemma) Small masses of vegetative tissue that form on the upper surface of a thallus that can be dispersed for asexual, vegetative reproduction.

Generative cell The cell in a pollen grain that divides to become two sperm cells.

Genus The taxonomic rank below family and above species.

Germination The resumption of growth by a dormant spore, seed, or pollen grain.

Gills The plates on the underside of the caps of basidiomycete fungi.

Green algae The group of algae that gave rise to plants.

Gymnosperms Plants that reproduce using cones and bear exposed "naked" seeds (that is, not contained in fruits).

Gynoecium The collective term for all of the carpels in a flower.

Haploid Having only one set of chromosomes.

Herb A nonwoody plant.

Herbarium A place where preserved plant materials are stored for study.

Herbarium sheets Pieces of paper upon which pressed, dried plant specimens are mounted; they typically provide information such as the plant's scientific name and where and by whom it was collected.

Heterospory The production of two types of spores: microspores that give rise to male gametophytes and megaspores that give rise to female gametophytes.

Heterotrophs Organisms that must obtain energy by consuming other animals.

Holdfast Part of the algal body that attaches the alga to the substrate.

Homospory The production of one type of spore.

Hydroids Primitive water-conducting cells in mosses.

Hyphae (singular: hypha) Fungal filaments.

Indusium A protective flap that covers immature sori in ferns.

Inflorescence A cluster of flowers close to one another on a stem.

Integuments Protective outer layers that surround the sporangium of an ovule.

International Botanical Congress The governing body that establishes rules on botanical nomenclature.

International Code of Botanical Nomenclature (ICBN) The internationally agreed upon set of rules that govern the naming of plants.

Kingdom The taxonomic level below domain.

Leaves The flattened photosynthetic parts of a vascular plant.

Legumes Fruits that are characteristic of the bean family.

Leptoids Primitive food-conducting cells of mosses.

Lichen Organism that results from a symbiotic relationship between a fungus and an alga.

Lignin A component of the hardened cell wall in xylem, the primary component of wood.

Mating types Genetically different forms of algae capable of sexually reproducing with one another.

Glossary

Megaphylls Large leaves with many veins.

Megasporangium A structure in which spores are formed that will become a female gametophyte.

Megaspores Spores that give rise to the female gametophyte.

Meiosis The process whereby a diploid cell divides twice to produce four haploid cells.

Microphylls Generally small leaves with a single vein.

Micropyle A small opening in the seed plant ovule that the pollen tube enters for fertilization.

Microsporangium A structure in which spores that will become the male gametophyte are formed.

Microspores Spores that will give rise to the male gametophyte.

Mitosis The process whereby one diploid cell divides once to produce two identical diploid cells.

Monocots Angiosperms whose seeds contain a single cotyledon.

Monophyletic group A group composed of an ancestor and all of its descendants.

Morphology The study of the structure and form of organisms.

Mutualism A relationship that benefits both participating species.

Mycelium A mass of fungal hyphae.

Mycology The study of fungi.

Mycorrhizae Fungi that engage in symbiotic relationships with plant roots.

Natural classification system A system of grouping organisms that is based on their evolutionary relationships.

Natural selection A process of evolutionary change that occurs when genetic change produces individuals with greater reproductive success or greater survival.

Nitrogen fixing Process of converting nitrogen from a gas into a biologically useful form.

Nucleus Part of a eukaryotic cell that contains DNA.

Organelles Small, specialized structures within a eukaryotic cell.

Ovary The enlarged, seed-producing portion of a flower that, after fertilization, becomes a fruit.

Ovules Structures in seed plants that contain the female gametophyte.

Pedicel The stalk of a single flower in an inflorescence.

Peduncle The stalk of a solitary flower or inflorescence.

Perennials Plants that live for more than two years and typically reproduce repeatedly throughout their lives.

Petals The parts of a flower that are often brightly colored.

Petiole The stalk of a leaf that attaches the blade to the stem.

Phloem Tissue that transports sugars and other products of photosynthesis throughout the vascular plant body

Photoautotrophs Organisms capable of forming their own energy resources through photosynthesis.

Photosynthesis The process through which plants convert the energy in light into sugars and oxygen.

Phycobilins A group of reddish pigments found in red algae and blue-green algae.

Phycology The study of algae.

Phylocode A nomenclature system that has no hierarchical rankings and recognizes only clades.

Phylogeny The evolutionary history of an organism or group of organisms.

Pistil The collective term for the female parts of a flower.

Plantae The plant kingdom.

Polar nuclei The two nuclei located in the female gametophyte of angiosperms, which fuse with a sperm to form the endosperm.

Pollen A structure containing the sperm cells in angiosperms and gymnosperms.

Pollen tube A tube that develops from a pollen grain and carries the sperm to the egg.

Glossary

Pollination In angiosperms, the transfer of pollen from an anther to a stigma; in gymnosperms, the transfer of pollen from a male cone to a female cone.

Pollinators Organisms that transfer pollen from one flower to another.

Polynomial An outdated system of assigning a name of multiple parts to one species.

Progymnosperms A group of now-extinct plants that gave rise to gymnosperms.

Prokaryotic Organisms whose cells lack nuclei and membrane-bound organelles.

Protista The protist kingdom.

Protonemas The threadlike gametophytes of some nonvascular and seed-less vascular plants.

Receptacle The structure in a flower where the floral whorls attach.

Red algae A large group of mostly multicellular marine species that get their characteristic color from reddish pigments in their cells.

Rhizoids Primitive structures that nonvascular plants use to attach to their substrate.

Scientific name The official name for a plant, consisting of two words: the genus and the species.

Seed ferns Extinct group of fernlike gymnosperms that produced seeds at the tips of their leaves.

Seedless vascular plants A group of plants that includes ferns, horsetails, and club mosses that have vascular tissue, but reproduce by spores.

Seeds Fertilized plant ovules consisting of an embryo and its food source.

Sepals Leaflike outermost structures of a flower.

Seta Stalk that supports the capsule in the moss sporophyte.

Sexual reproduction The formation of offspring by combining egg and sperm.

Shrub A woody plant that produces several stems and is shorter that a tree.

Sori The clusters of sporangia on the leaf surface of a fern.

Spadix A special type of inflorescence found in some angiosperms.

Species A particular type of organism that can be differentiated from other types of organisms; all members of a species have the ability to inter-breed and share a common evolutionary history.

Sporangium A spore-producing structure.

Spores Single-celled reproductive structures in bryophytes and seedless vascular plants that are capable of developing into a plant.

Spore mother cells Cells in a plant's reproductive tissue that undergo meiosis to form haploid spores.

Sporophylls Modified leaves that bear sporangia.

Sporophyte The diploid, spore-producing stage in the alternation of generations.

Stamens The male parts of flowers, composed of anther and filament, that produce pollen.

Sterile jacket A layer of cells that surrounds the gamete-producing arche-gonium and antheridium.

Sterile whorls The parts of a flower (sepals and petals) that do not pro-duce gametes.

Stigma The receptive potion of the carpel upon which pollen grains germinate.

Stipe The stem of a brown algae body.

Stomata Openings in the leaf surface that allow a plant to take in and release gasses.

Strobilus (plural: strobili) A cone-like cluster of spore-bearing leaves.

Style The carpel tissue that connects the stigma to the ovary; pollen tubes grow through the style to reach the ovary.

Succession A series of predictable, cumulative changes in the composition and characteristics of a plant community following disturbance.

Symbiotic A relationship in which both individuals benefit from the inter-action and are harmed when they are not together.

Synergids Two cells located near the egg in the female gametophyte of angiosperms.

Glossary

Systematics The study of evolutionary relationships and diversity.

Taxon A group of organisms at any hierarchical level, such as kingdom or species.

Taxonomy The science of classifying organisms.

Tepal Sepals and petals that cannot be distinguished from one another.

Thallus A simple vegetative body, undifferentiated into root, stem, or leaf.

Three-merous Having three sepals, three petals, three or six stamens, and a pistil with three carpels.

Tracheids Elongated, thick-walled xylem cells found in most vascular plants that conduct and support.

Tracheophytes Another name for vascular plants.

Tree A woody plant with a single trunk.

Tube cell The cell that develops into the pollen tube in the male gametophytes of seed plants.

Vascular bundles Strands of tissue that contain xylem and phloem.

Vascular cambium A localized area of cell division and growth in plants that produces new vascular tissue and contributes to increased diameter of woody stems.

Vascular tissue Tissue used to transport water and minerals throughout the plant body.

Vegetative reproduction Asexual reproduction through the breaking off of a part of the plant body to produce a new plant.

Whorl An arrangement of three or more floral parts or leaves in a circle.

Xylem Tissue that transports water and minerals throughout the vascular plant body

Zygospore A dormant spore formed in zygomycetes and some algae.

Zygote The diploid first cell of a sporophyte that results from the joining of sperm and egg at fertilization.

Attenborough, D. *The Private Life of Plants.* Princeton, N.J.: Princeton University Press, 1995.

Botany Online, The Internet Hypertextbook. Available online at http://www.biologie.uni-hamburg.de/b-online/e00/contents.htm.

Burleigh, J.G., and S. Mathews. "Phylogenetic Signal in Nucleotide Data From Seed Plants: Implications for Resolving the Seed Plant Tree of Life." *American Journal of Botany* 91 (2004): pp. 1599–1614.

Camera Wire Service. "Really Old Growth, Prehistoric Pines Found in Australia." *Boulder Daily Camera* (1994): 1A, 8A.

Chase, M.W. "Monocot Relationships: An Overview." *American Journal of Botany* 91 (2004): pp. 1645–1656.

Cowen, R. *History of Life.* Malden, Mass.: Blackwell Publishing, 2005.

Crane, P.R., P. Herendeen, and E.M. Friis. "Fossils and Plant Phylogeny." *American Journal of Botany* 91 (2004): pp. 1683–1699.

Crepet, W. "Early Bloomers." *Natural History* 108 (1999): pp. 40–41.

Darwin, Charles R. "Letter to J.D. Hooker, July 22nd 1879," *in* Darwin, F. and A.C. Seward, (eds.) *More Letters of Charles Darwin: A Record of His Work in a Series of Hitherto Unpublished Papers. Vol II.* pp. 20–21. London, UK: John Murray, 1903.

Futuyma, D.J. *Evolution.* Sunderland, Mass.: Sinauer, 2005.

Hummel, A.W. "The Printed Herbal of 1249 A.D." *Isis* 33 (1941): pp. 439–442.

Judd, W.S., C.S. Campbell, E.A. Kellogg, P.F. Stevens, and M.J. Donoghue. *Plant Systematics A Phylogenetic Approach. 2nd Ed.* Sunderland, Mass.: Sinauer, 2002.

Klesius, M. "The Big Bloom." *National Geographic* 202 (2002): pp. 102–121.

Levetin, E., K. McMahon. *Plants and Society. 2nd Ed.* Boston: WCB McGraw-Hill, 1999.

Lewis, L.A., R.M. McCourt. "Green Algae and the Origin of Land Plants." *American Journal of Botany* 91 (2004): pp. 1535–1556.

Linnaeus, C. "Genera Planatarum." 1787. *in* Baigrie, B.S. *Scientific Revolutions. Primary texts in the History of Science.* Upper Saddle River, N.J.: Pearson Prentice Hall, 2004.

Lutzoni, F., F. Kauff, C.J. Cox, D. McLaughlin, G. Celio, B. Dentinger, et al. "Assembling the Fungal Tree of Life: Progress, Classification, and Evolution of Subcellular Traits." *American Journal of Botany* 91 (2004): pp. 1446–1480.

Bibliography

Milius, S. "Should We Junk Linnaeus?" *Science News* 156 (1999): pp. 268–270.

Niklas, K. "What's So Special About Flowers?" *Natural History* 108 (1999): pp. 42–45.

Palmer, J.D., D.E. Soltis, and M.W. Chase. "The Plant Tree of Life: An Overview and Some Points of View." *American Journal of Botany* 91 (2004): pp. 1437–1445.

Pryer, K.M., E. Schuettpelz, P.G. Wolf, H. Schneider, A.R. Smith, and R. Cranfill. "Phylogeny And Evolution Of The Ferns (Monilophytes) With A Focus On The Early Leptosporangiate Divergences." *American Journal of Botany* 91 (2004): pp. 1582–1598.

Raven, P.H., G.B. Johnson, J.B. Losos, and S.R. Singer. *Biology.* 7th Ed. New York: McGraw-Hill, 2005.

Regal, P.J. "Ecology and Evolution of Flowering Plant Dominance." *Science* 196 (1977): pp. 622–629.

Rost, T.L., M.G. Barbour, C.R. Stocking, and T.M. Murphy. *Plant Biology.* 2nd Ed. Toronto, Canada: Thompson Brooks/Cole, 2006.

Sanderson, M.J., J.L. Thorne, N. Wikstrom, and K. Bremer. "Molecular Evidence On Plant Divergence Times." *American Journal of Botany* 91 (2004): pp. 1656–1665.

Savage, J.M. "Systematics and the Biodiversity Crisis." *BioScience* 45 (1995): pp. 673–679

Shaw, J., and K. Renzaglia. "Phylogeny and Diversification of the Bryophytes." *American Journal of Botany* 91 (2004): pp. 1557–1581.

Simpson, B.B., and J. Cracraft. "Systematics: The Science of Biodiversity." *BioScience* 45 (1995): pp. 670–672

Soltis, P.S., and D.E. Soltis. "The Origin and Diversification of Angiosperms." *American Journal of Botany* 91 (2004): pp. 1614–1626.

Tree of Life Web Project. Available online at http://tolweb.org/tree/phylogeny.html

Uno, G., R. Storey, and R. Moore. *Principles of Botany.* New York: McGraw-Hill Higher Education, 2001.

USDA PLANTS database. Available online at http://www.plants.usda.gov

Withgott, J. "Is It 'So Long Linnaeus'?" *BioScience* 50 (2000): pp. 646–651.

Zimmer, E.A., Y.-L. Qio, P.K. Endress, and E.M. Friss. "Current Perspectives On Basal Angiosperms." *International Journal of Plant Sciences* 161 Supplement (2000): S1–S2.

Anderson, Edgar. *Plants, Man and Life.* St. Louis, Mo.: Missouri Botanical Garden, 1997.

Baskin, Carol C. and Jerry M. Baskin. *Seeds. Ecology, Biogeography, and Evolution of Dormancy and Germination.* San Diego, Calif.: Academic Press, 1998.

Berlin, Brent, Dennis E. Breedlove, and Peter H. Raven. "Folk Taxonomies and Biological Classification." *Science* 154 (1966): pp. 273–275.

Blackmore, Stephen and Elizabeth Tootill, eds. *The Facts on File Dictionary of Botany.* New York: Facts on File, Inc., 1984.

Durrell, Gerald. *A Practical Guide for the Amateur Naturalist.* London, U.K.: Alfred A. Knopf, Inc., 1982.

Erickson, Jon. *A History of Life on Earth: Understanding Our Planet's Past.* New York: Facts on File, Inc., 1995.

Gould, Steven J. "Linnaeus's Luck?" *Natural History* (2000): pp. 18–25, pp. 66–69, pp. 74–76.

Miller, G. Tyler, Jr. *Essentials of Ecology.* Pacific Grove, Calif.: Brooks Cole, 2005.

Miller, Douglass R. and Amy Y. Rossman. "Systematics, Biodiversity, and Agriculture." *BioScience* 45 (1995): pp. 680–686.

Morell, Virginia. "The Variety of Life." *National Geographic* 195 (1999): pp. 6–31.

Smith, James P., Jr. *Vascular Plant Families.* Eureka, Calif.: Mad River Press, Inc., 1977.

Wilson, Edward O. "Biodiversity: Challenge, Science, Opportunity." *American Zoologist* 32 (1992): pp. 1–7.

Wilson, Edward O. *The Future of Life.* New York: Vintage Books, 2002.

Young, Paul. *The Botany Coloring Book.* New York: HarperCollins Publishers, 1982.

Web Sites
American Bryological and Lichenological Society
http://www.unomaha.edu/~abls/

American Fern Society
http://amerfernsoc.org

Angiosperm Phylogeny Website
http://www.mobot.org/MOBOT/research/APweb/

Further Reading

Botanical Society of America
http://www.botany.org

Ecological Society of America
http://www.esa.org

Flora of North America
http://www.fna.org/FNA/

Fungi Online
http://nt.ars-grin.gov/sbmlweb/fungi/index.cfm

Mycological Society of America
http://www.msafungi.org

National Biological Information Infrastructure
http://www.nbii.gov/disciplines/botany/

New York Botanical Garden Herbarium
http://www.nybg.org/bsci/herb/

Paleobotanical Section of the Botanical Society of America
http://www.dartmouth.edu/~daghlian/paleo/

Phylogeny of Life
http://www.ucmp.berkeley.edu/exhibit/phylogeny.html

Royal Botanic Gardens, Kew
http://www.rbgkew.org.uk

Smithsonian National Museum of Natural History
http://www.mnh.si.edu

Wayne's Word Online Textbook of Natural History
http://waynesword.palomar.edu

page:

2-3: Igor Karon / www.shutter stock.com

5: (a) Winthrop Brookhouse / www.shutterstock.com, (b) Aleksander Bolbot / www. shutterstock.com, (c) Rodney Mehring / www.shutterstock .com, (d) Anette Linnea Rasmussen / www.shutter stock.com

14: © Infobase Publishing

16: © Infobase Publishing

18-19: Merryl McNaughton / www. shutterstock.com

21: Georgette Douwma / Photo Researchers, Inc.

24: Library of Congress [LC-USZ62-11324]

28: © Infobase Publishing

30: © Scott Camazine / PhototakeUSA.com

32-33: Michael Stevens / Shutterstock.com

35: Josh Meyer / Shutterstock.com

38: (a) Chris Hellyar / Shutterstock.com, (b) Keith Weller, (c) Romeo Mihulic / www.shutterstock.com, (d) Andre Nantel / www.shutter stock.com

39: © Infobase Publishing

43: Inga Spence / Visuals Unlimited

48: Robert and Jean Pollock / Visuals Unlimited

50-51: Zavodskov Anatoliy Nikolaevich / www.shutter stock.com

53: Anne Kitzman / www.shutter stock.com

55: Amygdala imagery / www. shutterstock.com

56: David T. Roberts / Natures Images / PhotoResearchers, Inc.

58: Kathy Merrifield / Photo Researchers, Inc.

62-63: C. Salisbury / www.shutter stock.com

67: Perennou Nuridsany / Photo Researchers, Inc.

69: Michael P. Gadomski / Photo Researchers, Inc.

72: Peter Hansen / www.shutter stock.com

74: © Infobase Publishing

75: © Jacques Jangoux / Visuals Unlimited

78-79: Tian Rencelj / www.shutter stock.com

80: M. Philip Kahl / Photo Researchers, Inc.

81: Peter Blazek / www.shutter stock.com

83: (a) Susan McKenzie / www. shutterstock.com, (b) Vladimir Ivanov / www.shut terstock.com, (c) Chandral Photo / www.shutterstock .com, (d) Kathy Merrifield / Photo Researchers, Inc.

88: Kathy Merrifield / Photo Researchers, Inc.

90: Gilbert S. Grant / Photo Researchers, Inc.

94-95: Mark Grenier / www.shutter stock.com

97: © Infobase Publishing

99: © Infobase Publishing

103: (a) © Biodisc / Visuals Unlimited, (b) © Dr. John D. Cunningham / Visuals Unlimited

Picture Credits

Index

Index

Index

About the Authors

J. Phil Gibson holds degrees in botany from Oklahoma State University (B.S.) and the University of Georgia (M.S.), and a Ph.D. in environmental population and organismic biology from the University of Colorado. He is currently an associate professor in the Department of Botany and Microbiology and the Department of Zoology at the University of Oklahoma. His research investigates the ecology and evolution of plant reproductive systems. He also conducts conservation-focused research on tree species. He has published a variety of research papers and presented his work at scientific conferences. Gibson is a member of the Project Kaleidoscope Faculty for the 21st Century in recognition of his efforts to improve undergraduate science education. He is also an active member of the Botanical Society of America and the Association of Southeastern Biologists.

Terri R. Gibson holds a bachelor's degree in zoology from the University of Georgia. She has worked as a scientific illustrator and also as a research assistant studying, among other things, plant population genetics, plant morphology, and human immunodeficiency virus (HIV). She is currently pursuing a career in children's literature.